New Therapies to Prevent or Cure Auditory
Disorders

Sylvie Pucheu • Kelly E. Radziwon
Richard Salvi
Editors

New Therapies to Prevent or Cure Auditory Disorders

 Springer

Editors
Sylvie Pucheu
CILcare
Advanced Solution for Drug Development
in Hearing Disorder
Montpellier, France

Kelly E. Radziwon
Center for Hearing and Deafness
University at Buffalo
Buffalo, NY, USA

Richard Salvi
Center for Hearing and Deafness
University at Buffalo
Buffalo, NY, USA

ISBN 978-3-030-40415-4   ISBN 978-3-030-40413-0   (eBook)
https://doi.org/10.1007/978-3-030-40413-0

This Springer imprint is published by the registered company Springer Nature Switzerland AG
The registered company address is: Gewerbestrasse 11, 6330 Cham, Switzerland

Introduction

Hearing disorders affect more than 15% of people worldwide. One out of three people over 65 years of age has disabling hearing loss, with 80% of young people at significant risk of hearing impairment. The World Health Organization (WHO) estimates that by 2050, 10% of the world's population will have disabling hearing loss. Currently, sensorineural hearing loss (SNHL) is the most common type of hearing disorder, accounting for almost 90% of reported hearing loss. SNHL is characterized by the loss of sensory hair cells in the cochlea and/or neural damage in the auditory pathway. Since hair cells cannot regenerate in the human cochlea, SNHL is permanent. Multifactorial etiologies for SNHL include aging, noisy lifestyles and work environments, infections, and more than 300 ototoxic drugs on the market, including the commonly used cancer drug cisplatin. This volume provides an up-to-date resource on the mechanisms underlying hearing disorders, including related pathologies such as tinnitus, a phantom auditory sensation, and hyperacusis, a loudness intolerance disorder, and provides potential strategies for ameliorating auditory dysfunction.

The first chapter by Shinichi Someya et al. provides an overview of the current literature on interventions for age-related hearing loss (AHL), also known as presbycusis, with a particular emphasis on calorie restriction (CR), a lifestyle-based intervention. Chapter 1 details the mechanisms underlying the beneficial effects of CR on auditory function in laboratory animals and humans. Chapter 2 by Celia Escabi et al. describes the anatomical, physiological, and perceptual consequences of noise trauma, as well as emerging treatments to alter the biological mechanisms, broadly classified as metabolic (oxidative stress, excitotoxicity, or apoptosis) or mechanical, that normally lead to injury and death of cells in the organ of Corti, spiral ganglion, and/or the lateral wall.

Chapter 3 by Chaitanya Mamillapalli et al. outlines the effects of ototoxic drugs, primarily cisplatin, cyclodextrins, and aminoglycosides, on the auditory system, and potential strategies to mitigate their impact on hearing. Chapter 4 by Muhammad Waqas and Renjie Chai details the most promising therapeutic approaches to regenerate cochlear hair cells (HCs) and spiral ganglion neurons (SGNs) including gene therapy and stem cell therapy. Chapter 5 by Bohua Hu and

Celia Zhang describes cochlear inflammatory activities in acute and chronic stress conditions and suggests that controlling the cochlear immune state could offer protection against pathogenesis.

In Chap. 6, Kelly E. Radziwon et al. provide a review on hyperacusis, including its definition and various manifestations, and describe the available behavioral procedures used to induce and study hyperacusis in preclinical animal models. The final chapter (Chap. 7) by Amandine Laboulais et al. introduces two robust measures of hearing loss and tinnitus in preclinical models involving a new automatic counting method to quantify hair cell loss and an objective imaging method developed during a collaborative project between CILcare, KeenEye, and Charles Coulomb Laboratory. Ultimately, the goal of each chapter is to provide relevant insight into current models of hearing disorders, including their underlying mechanisms, which will help pharmaceutical and biotechnology companies develop therapies to treat and prevent these hearing impairments.

Buffalo, NY, USA   Richard Salvi
CILcare, Advanced Solution for Drug Development   Sylvie Pucheu
in Hearing Disorder, Montpellier, France
Buffalo, NY, USA   Kelly E. Radziwon

Contents

Lifestyle Intervention to Prevent Age-Related Hearing Loss: Calorie Restriction

Shinichi Someya, Christina Rothenberger, and Mi-Jung Kim

Abbreviations

$\bullet O_2^-$	Superoxide
$\bullet OH$	Hydroxyl radical
ABR	Auditory brainstem response
AHL	Age-related hearing loss
CAT	Catalase
Complex I	NADH dehydrogenase
Complex III	Ubiquinone–cytochrome c reductase
CR	Calorie restriction
GPX1	Glutathione peroxidase 1
GSH	Reduced glutathione
GSR	Glutathione reductase
GSSG	Oxidized glutathione
GSTM1	Glutathione *S*-transferase mu 1
GSTP1	Glutathione *S*-transferase pi 1
GSTT1	Glutathione *S*-transferase theta 1
H_2O_2	Hydrogen peroxide
IHC	Inner hair cells
mtDNA	Mitochondrial DNA
NIA	National Institute on Aging
NIHL	Noise-induced hearing loss
OHC	Outer hair cells
PRDX3	Peroxiredoxin 3
ROS	Reactive oxygen species
SGN	Spiral ganglion neuron
SOD1	Superoxide dismutase 1
SOD2	Superoxide dismutase 2
SV	Stria vascularis
TXNRD	Thioredoxin reductase
UW	University of Wisconsin–Madison

S. Someya (✉) · C. Rothenberger · M.-J. Kim
Department of Aging and Geriatric Research, University of Florida, Gainesville, FL, USA
e-mail: someya@ufl.edu

© Springer Nature Switzerland AG 2020
S. Pucheu et al. (eds.), *New Therapies to Prevent or Cure Auditory Disorders*,
https://doi.org/10.1007/978-3-030-40413-0_1

1

1   Introduction

An accumulating body of evidence indicates that reducing calorie intake or calorie restriction (CR) extends life span in diverse species and delays the onset of a variety of age-related diseases, including hypertension, diabetes, cancer, and cardiovascular disease in laboratory animals. In humans, CR reduces the incidence of obesity and levels of cholesterol, blood pressure, oxidative stress, and inflammation, and increases insulin resistance [1]. In general, CR results in weight loss, which in turn increases health span of experimental animals [1–3]. The anti-aging effects of CR require significant reduction of body weight through reducing food consumption. This idea is supported by the observation that food shortages during World War I in European countries were associated with a sharp decrease in coronary heart disease mortality, which increased again after the war ended [4]. Another study among Spanish nursing home residents undergoing long-term alternate day feeding regimen also demonstrated decreased morbidity and mortality [5]. In Japan, inhabitants of Okinawa island, who ate ~30% fewer calories than the rest of Japanese residents, had ~35% lower rates of cardiovascular disease and cancer mortality than the average Japanese population and had one of the highest numbers of centenarians in the world [6]. However, due to Westernization of their diet and/or excessive calorie intake, the life expectancy at birth for men in Okinawa is now no higher than the national average in Japan, reflecting increased mortality due to the increase in heart disease and cerebrovascular disease [7].

Dietary weight loss can also impact physical function by improving muscle quality and reducing intramuscular adipose tissue [8, 9]. Consistent with these reports, age-related hearing loss (AHL), a common feature of aging, is also associated with decreased physical function among older adults [10, 11], while long-term exercise reduces body weight and slows the progression of AHL in mice [12]. Increasing evidence indicates that CR has beneficial effects on hearing in laboratory animals. This chapter reviews the current literature on interventions for AHL, particularly focusing on a lifestyle-based intervention, CR, and what has been learned about the mechanisms underlying the beneficial effects of CR on auditory function in laboratory animals and humans.

2   Effects of Age on Auditory Function

Hearing gradually declines with age in laboratory animals and humans [13]. Hearing loss is the third most prevalent chronic health condition affecting older adults and AHL, also known as presbycusis, is the most common form of hearing impairment [13–15]. WHO estimates that one-third of persons over 65 years are affected by disabling hearing loss [16]. Worldwide, approximately 466 million people suffer from hearing impairment and this number is expected to rise to 630 million by 2030 and over 900 million in 2050 [16]. Approximately, 30 million or 13% of Americans

12 years and older have bilateral hearing loss in 2001–2008, and this number rises to 48 million when individuals with unilateral hearing loss are included [17]. Because the prevalence of AHL is expected to rise dramatically as the world's population ages, AHL will become a major social and health care problem for which there are no established cures or treatments.

AHL is characterized by poor speech understanding particularly in noise, impaired temporal resolution, and central auditory processing deficits [13–15]. AHL is also associated with dementia [18]. Despite its name, AHL is likely a multifactorial condition resulting from the interaction of numerous causes including aging, exposure to noise and ototoxic chemicals, genetics, epigenetic variables, comorbidities, and lifestyle [13]. The major sites of age-related cochlear pathology include inner hair cells (IHC), outer hair cells (OHC), spiral ganglion neurons (SGN), stria vascularis (SV), and synaptic loss [13, 14, 19]. The IHC are the sensory receptors that relay their electrical response to the central auditory system through the SGN [20]. Postmitotic hair cells and SGN are particularly susceptible to injury from a combination of noise exposure, ototoxic chemicals, and oxidative damage [13, 14]. The blood vessels coursing through the cochlea are essential for transporting oxygen and nutrients such as glucose into the cochlea [13]. Thus, age-related degeneration of the hair cells, SGN, and cochlear vasculature will disrupt auditory function and lead to permanent hearing impairment. Schuknecht [21] identified four major categories of presbycusis, namely (1) sensory (hair cell loss), (2) neural (SGN loss), (3) strial (SV degeneration), and (4) cochlear conductive based on correlations between cochlear pathology and the audiogram. Numerous studies indicate that age-related degeneration of the stria vascularis is one of the most prominent features of AHL in animals and humans [14, 22]. Indeed, AHL is often associated with significant loss of strial capillaries in the lateral wall of the gerbil cochlea [23], suggesting that an age-related decline in blood flow, oxygen, and nutrient transport to the cochlea plays a key role in the development of AHL.

3 Effects of Calorie Restriction on Aging

An accumulating body of evidence indicates that reducing calorie intake, i.e., CR, is the only intervention that can slow the rate of aging in diverse species. Indeed, CR extends the life span of rodents, nonhuman primates, birds, protozoans, water fleas, spiders, and guppies [1, 3]. CR also delays the onset of a variety of age-related diseases, including hypertension, diabetes, cancer, cardiovascular disease, Parkinson's disease, Alzheimer's disease, cataracts, and AHL in rodents [1–3, 24]. In rhesus monkeys, CR results in signs of improved health, including reduced incidence of obesity, diabetes, and tumors [25, 26]. In humans, CR reduces the incidence of obesity, levels of cholesterol, blood pressure, oxidative stress, and inflammation, and increases insulin resistance [1]. At the molecular level, CR results in a reduction in the levels of oxidative protein damage in the brain, heart, and liver of aged mice [27].

Maximum life span is thought to be increased by reducing the rate of aging, while the average life span is primarily influenced by improving environmental conditions [1–3]. Consistent with this idea, the average life span of humans has dramatically increased as a result of improved diet and environmental health conditions, whereas the maximum life span has remained largely unchanged. Herein, we review the current literature on the beneficial effects of CR on AHL in laboratory animals and humans.

3.1 Types of Calorie Restriction Regimens

Two different dietary feeding regimens of CR have been widely used because of their feasibility and ability to reliably slow aging in laboratory animals [28]. In an "every day feeding" regimen, animals receive food daily, but are limited to a specified amount which is usually 30–50% less than the ad libitum consumption by the control group. In an "every other day feeding" or "intermittent fasting" regimen, animals are deprived of food for a full day, every other day, and are fed ad libitum on the intervening days. In animal CR studies, the species are generally housed individually since it allows for the food intake of each animal to be controlled with great accuracy [29].

3.2 Effects of Calorie Restriction on Hearing Loss in Laboratory Animals

Increasing evidence indicates that CR has beneficial effects on auditory function in laboratory rodents. The C57BL/6 (B6) mouse strain has been used extensively to study aging and this murine species responds to CR with a robust extension of life span [2, 30]. B6 mice also display the classic pattern of AHL, which includes severe to profound loss of hearing beginning at the high frequencies and loss of high-frequency, basal turn cochlear hair cells by 10–12 months of age [31, 32]. To investigate the effects of CR on AHL in B6 mice, Someya et al. [24] restricted the calorie intake of male B6 mice to 75% of that given to control mice starting at 2 months of age using an every other day feeding regime. Both control and CR mice were fed diets only on Mondays, Wednesdays, and Fridays. CR mice received 63 kcal/week of the diet, while control mice received 84 kcal/week of the diet. The control mice were not fed ad libitum daily to avoid large individual variations among the caloric intake of the control group and obesity. These animals were housed individually until they reached 12 months of age or for a 13-month period. Auditory brainstem response (ABR) testing was used to assess hearing sensitivity in these animals. Middle-age control diet mice displayed profound hearing loss at 4, 8, and 16 kHz, whereas the thresholds of age-matched CR mice were significantly lower than those age-matched controls. However, the thresholds of CR and controls were not significantly different from one another at 4 months of age at all the frequencies measured. These results indicate that CR can slow the onset of AHL that would otherwise occur in male B6 mice by 15 months of age. CR mice weighed 25% less than the controls and displayed only minor cochlear degeneration compared to age-matched control diet mice. Willott et al. [33] also investigated the effects of CR on AHL in B6 mice. Male B6 mice were calorie restricted to 70% of the control intake. Animals were housed four per cage for a 22-month period from 1 month of age until 23 months of age and fed the CR or control measured amount of food daily. Total amounts of all diet components except starch and corn oil were the same. ABR testing was used to monitor hearing sensitivity and the progression of AHL. Consistent with the report by Someya and coworkers [24], CR significantly reduced SGN loss compared to controls. However, no ABR thresholds were measureable at 4, 8, 16, 24, and 32 kHz in both the control and CR mice at 23 months of age, indicating that all the mice displayed complete to nearly complete hearing loss. This may be due to the fact that the time at which hearing was measured in these B6 (23 months of age) was too late for CR to be beneficial since the B6 strains develop severe to complete hearing loss by 10–12 months of age [31, 32]. Summaries of studies demonstrating beneficial effects of CR on AHL in laboratory animals are given in Table 1.

The CBA/J (CBA) mouse strain is long-lived (mean life span = 679 days) [39], has normal hearing until later in life, and displays AHL by 18–20 months of age [32, 34]. Sweet and coworkers [34] investigated the effects of CR on CBA mice. CR animals were fed standard lab chow ad libitum every other day (on Mondays, Wednesdays, and Fridays), while control mice were fed ad libitum daily from 2 until 10 months of age or for an 8-month period (early-onset CR) or from 10 until

Table 1 Studies demonstrating beneficial effects of CR on hearing in laboratory animals

Species (strain)	Sex	CR method	Beneficial effects (age at test)	Reference
Mouse				
C57BL/6	Male	Every other day	Lower ABR thresholds (15 months old)	[24]
		25% CR		
CBA/J	NA	Every other day	Lower ABR thresholds (27 months old)	[34]
AU/Ss	NA	Every other day	Lower ABR thresholds (18 months old)	[35]
Rat				
Fischer 344	NA	Every day	Lower ABR thresholds (24–25 months old)	[36]
		30% CR		
Sprague–Dawley	Female	Every day	Reduced degeneration of SV (30 months old)	[37]
		30% CR		
Monkey				
Rhesus monkey	Male	Every day	Larger ABR wave amplitudes (11–23 years old)	[38]
	Female	30% CR		

27 months of age for a 17-month period (middle age-onset CR). All mice were housed with one to three per cage and ABR testing was used to assess hearing sensitivity. Both control and CR mice from the early-onset CR study displayed similar ABR threshold elevations at 4, 8, 16, 32, and 64 kHz at 10 months of age. However, in the middle-age-onset CR study, the mean ABR thresholds from CR mice were significantly lower than those of control mice at 8, 16, and 32 kHz at 27 months of age (Table 1), indicating that middle-age onset CR can delay the onset of AHL in CBA mice.

The AU/Ss (AU) mouse strain has normal hearing in early life, but develops gradual hearing loss in the second half of its life. The AKR strain is a short-lived mouse strain (mean life span = 323 days) [39] and develops early onset of hearing loss by 1–2 months of age [35]. Henry [35] investigated the effects of CR on AHL in AU and AKR mice. CR animals were fed standard lab chow ad libitum every other day on Mondays, Wednesdays, and Fridays, while control mice were fed ad libitum daily from 1 until 18 months of age for a 17-month period. All mice were housed one to three per cage and ABR testing was used to monitor the progression of AHL. At 18 months of age, the mean ABR thresholds of CR mice at 2, 4, 8, 32, and 64 kHz were significantly lower than those of controls, indicating that CR delayed the onset of AHL in AU mice (Table 1). CR mice also weighed 14% less than controls. In the CR study of AKR mice, CR animals were fed standard lab chow ad libitum every other day on Mondays, Wednesdays, and Fridays, while control mice were fed ad libitum daily from 2 until 4 months of age for a 2-month period. At 4 months of age, both control and CR mice displayed large threshold elevations at 2, 4, 32, and 64 kHz, and there were no differences in hearing threshold between control and CR animals at all the frequencies measured. This may be due to the fact that the onset of AHL in this strain occurs at a very young age (1–2 months of age), which may be too early for CR to be beneficial.

The Fischer 344 rat strain is long-lived with a median life span of 28–31 months [40] and develops age-related loss of hair cells and SGN cells and hearing loss by 18–20 months of age that begins in the high frequencies [41, 42]. In a study of CR in male Fischer 344 rats, animals were calorie restricted to 70% of the control intake from 2 months of age until 24–25 months of age [36]. Rats were housed individually. CR rats were fed 11.2 g of a standard rodent diet daily, while control rats were fed ad libitum. The Fischer 344 strain of rats displayed severe hearing loss by 20 months of age [36]. At 24–25 months of age, both control and CR rats displayed large ABR threshold elevations at 3, 6, 9, 12, and 18 kHz, but the mean ABR thresholds of CR rats were significantly lower than those of controls at all the frequencies measured. Old CR animals also had significantly less hair cell loss compared to age-matched controls suggesting that CR delays the onset of AHL in Fischer 344 rats by suppressing hair cell degeneration (Table 1).

The Sprague–Dawley rat strain also exhibits a loss of hair cells and spiral ganglion neurons that increases with age [43, 44] and severe age-related hearing loss by 18–20 months of age [45]. In female Sprague–Dawley rats, animals were calorie restricted to 70% of the control intake from 2 months of age until 30 months of age. Rats were group-housed three or four per cage and control rats were fed ad libitum [37]. In the CR groups, 73% of the CR rats displayed only modest degeneration of the SV, while age-matched control diet rats showed a marked thinning, cellular degeneration, and loss of cell processes in the SV (Table 1). The extent of loss of sensory hair cells was similar in both CR and control diet groups, while neither group showed a significant reduction in the number of SGN across their adult life span. These results indicate that CR delayed age-related degeneration of SV in Sprague–Dawley rats rather than delaying the degeneration of hair cells or SGN.

Rhesus monkeys are long-lived nonhuman primates with a maximum life span of 40 years [46] and display decreased auditory function by 25–31 years of age [46, 47]. Currently, there are two studies ongoing at the University of Wisconsin–Madison (UW) and the National Institute on Aging (NIA) designed to investigate the effects CR on life span and health span in rhesus monkeys [25, 26]. In the UW study, CR was maintained at 30% less than the calories consumed by the ad libitum group for a period of 3–9 years [25]. The animals were housed individually to minimize aggressive encounters and to control food access. All animals were daily fed pelleted, semi-purified diets, which contained 15% lactalbumin, 10% corn oil, and 65% carbohydrate in the form of sucrose and corn starch. At the time of the auditory tests, the ages of the monkeys were 11–23 years. ABR testing was used to monitor the progression of AHL [38]. The authors found that: (1) wave I amplitudes were larger for females and for younger monkeys than male monkeys, (2) amplitudes decreased in aging males, (3) wave IV amplitudes were larger for females than males, (4) wave IV amplitudes for CR females were larger than for female controls, whereas the amplitudes from control and CR males were not different, and (5) the mean ABR threshold (click stimuli) of CR male monkeys was lower than that of control males, but the difference was not statistically significant (Table 1). In general, these results were in agreement with the previous reports that women of virtually all ages demonstrate lower hearing thresholds, shorter auditory brainstem response (ABR) latencies, and larger ABR amplitudes than men [17, 48–54]. Together, these results indicate that CR delayed age-related decline in some components of auditory function in aged monkeys. In the NIA study, CR was maintained at 30% less than the calories consumed by the control group [55]. Dietary feeding and housing regimens were similar to those of the UW study. ABR testing was used to monitor the progression of AHL. In contrast to the UW study, no significant effects of CR were found on any ABR parameters examined. At the time of the hearing tests, the age range of the control diet monkeys was 13–32 years, while that of the CR monkeys was 13–36 years. Therefore, the lack of effect of CR may be due to the fact that some of the monkeys were still relatively young at the time of the tests and had not reached the age (~25 years) at which AHL begins to be clearly evident with hearing tests where CR is most likely to have an effect on auditory function.

4   Mechanisms Underlying the Beneficial Effects of CR on AHL

It is well documented that CR delays the development of a variety of age-related diseases and extends life span in a variety of species [1–3]. Despite numerous reports, the question still remains whether CR slows aging and other disease in long-living humans. This is primarily because the maximum life span of humans is 122+ years and it is impractical as well as unethical to conduct randomized, diet-controlled, long-term survival studies in humans. Similarly, the beneficial effects of CR against human AHL have not been tested. Nonetheless, increasing evidence from animal research suggests that CR likely has beneficial effects on human auditory function. Herein, we discuss the potential mechanisms underlying the beneficial effects of CR on the auditory system in mammals.

4.1   Oxidative Stress

It is well documented that CR reduces oxidative damage to DNA, proteins, and lipids [13–15, 56, 57] caused by reactive oxygen species (ROS) [2, 56, 58]. Mitochondria are a major source of ROS and ROS are known to damage key cellular components, including nuclear DNA, mitochondrial DNA (mtDNA), membranes, enzymes, and proteins [56]. Such damage accumulates with age, causes cell death, and eventually leads to tissue dysfunction. The majority of intracellular ROS are continuously generated as a by-product of mitochondrial respiratory metabolism during the generation of ATP [56, 58]. These ROS include superoxide ($\cdot O_2^-$) and the hydroxyl radical ($\cdot OH$), which are extremely reactive, and hydrogen peroxide (H_2O_2) that is freely diffusible and relatively long-lived [59, 60]. The production of superoxide as a by-product of mitochondrial respiration metabolism is thought to occur at two mitochondrial electron transport chain sites: Complex I (NADH dehydrogenase) and Complex III (ubiquinone–cytochrome c reductase). But under normal metabolic conditions, Complex III is thought to be the main site of ROS production (Fig. 1).

An elaborate defense system has evolved to reduce the damaging effects of ROS. This system includes the antioxidant enzymes such as superoxide dismutase (SOD), catalase (CAT), glutathione peroxidase (GPX), glutathione reductase (GSR), peroxiredoxin (PRDX), and thioredoxin reductase (TXNRD) [59, 60]. In mitochondria, SOD2 converts superoxide into hydrogen peroxide, which in turn is decomposed to water by GPX1 or PRDX3, thereby protecting key cell components such as nuclear DNA, mtDNA, proteins, and lipids from ROS-induced damage (Fig. 2). A variety of small-molecule antioxidants also scavenge ROS. These include glutathione, thioredoxin, carotenoids, ascorbate, and flavonoids [59–61]. Evidence indicates that the mitochondrial antioxidant defense systems do not keep pace with the age-related increase in ROS production, and that during aging, the balance

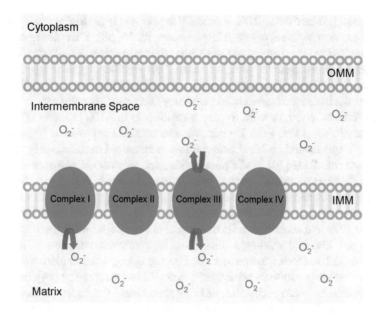

Fig. 1 The production of superoxide as a by-product of mitochondrial respiration metabolism. The production of superoxide as a by-product of mitochondrial respiration metabolism is thought to occur at two mitochondrial electron transport chain sites: Complex I (NADH dehydrogenase) and Complex III (ubiquinone–cytochrome c reductase). But under normal metabolic conditions, Complex III is thought to be the main site of ROS production

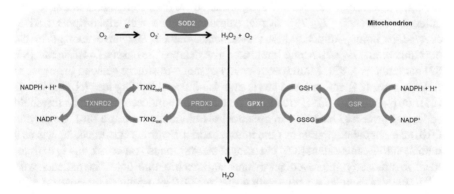

Fig. 2 Mitochondrial antioxidant defense system. In mitochondria, SOD2 converts superoxide into hydrogen peroxide, which in turn is decomposed to water by GPX1 or PRDX3, thereby protecting key cell components such as nuclear DNA, mtDNA, proteins, and lipids from ROS-induced damage

between the antioxidant defenses and ROS production shifts progressively toward a more pro-oxidant state [62]. Thus, the balance between ROS production and antioxidant defenses may determine the degree of ROS-induced oxidative damage.

Increased production of ROS is thought to play a role in driving the aging process and the development of age-related disease [2, 56, 58]. This idea is supported by the fact that overexpressing the mitochondrial antioxidant gene *Sod2* [63] or the mitochondrial iron regulator protein frataxin [64] significantly increases longevity in *Drosophila*. Similarly, overexpressing a mitochondrially targeted catalase gene (CAT) results in reduced age-related pathology and a moderate increase in life span in mice [65]. ROS are thought to play a major role in AHL [13, 66–68]. Several studies have shown that ROS are generated in cochleae exposed to high-intensity noise [69]. Age-related cochlear hair cell loss is enhanced in mice lacking the antioxidant enzyme *Sod1* [70], while mice lacking the antioxidant enzymes *Gpx1* [71] or *Sod1* [72] show enhanced susceptibility to noise-induced hearing loss. Moreover, oxidative protein damage increases with age in the cochleae of CBA/J mice [73]. Overexpression of mitochondrially targeted CAT also results in decreased loss of outer hair cells and inner hair cells in the cochlea and lower ABR thresholds compared to age-matched wild-type mice [74]. In agreement with the ABR results, cochlear oxidative DNA damage increased during aging, while oxidative damage to DNA was reduced by mitochondrially targeted CAT overexpression [74]. In humans, an association between AHL and *GSTT1* (glutathione *S*-transferase theta 1) and *GSTM1* (glutathione *S*-transferase mu 1) polymorphisms has been detected in the Finnish population [75]. A significant association has been observed between AHL and the *GSTM1* and the *GSTT1* polymorphisms in Hispanics [76], suggesting the glutathione detoxification system plays a protective role against AHL.

Numerous studies have also shown that supplementation with mitochondrial antioxidants have beneficial effects on AHL in humans and laboratory animals. For example, alpha-lipoic acid and *N*-acetylcysteine are thiol compounds linked to glutathione production [77, 78]. In rats, supplementation with alpha-lipoic acid or acetyl-*L*-carnitine, mitochondrial antioxidants, reduces oxidative damage in the heart and brain [79, 80]. Alpha-lipoic acid also delayed the onset of AHL in rats [81, 82] and mice [74, 83]. Moreover, *N*-acetyl cysteine treatment delayed the onset of AHL in mice [84] and reduced levels of noise-induced hearing loss (NIHL) in rats [85]. In guinea pigs, GSH treatment reduced noise-induced temporary threshold shifts and protected hair cells in the cochlea from noise exposure [86]. Coenzyme Q10, an essential component of the mitochondrial electron transfer chain, acts as a mitochondrial antioxidant [87]. Guastini et al. [88] found that coenzyme Q10 treatment significantly improved pure tone audiometric thresholds in patients with AHL. In mice, supplementation with coenzyme Q10 also delayed the onset of AHL at high frequencies [74].

In the mitochondrial matrix, SOD2 converts superoxide into hydrogen peroxide, which is then decomposed to water by GPX1 using glutathione as a substrate [59, 60] (Fig. 2). Oxidized glutathione (GSSG) is then regenerated to reduced glutathione (GSH) by GSR. Thus, GSR plays a critical role in maintaining the mitochondrial GSH/GSSG redox state and reducing mitochondrial ROS levels in cells. In healthy mitochondria from young mice, glutathione is found mostly in the reduced form [60] but during aging, the ratio of GSH/GSSG, or the glutathione redox state, declines in the mitochondria of brain, heart, eye, and testis of mice [62]. Because the GSH/GSSG redox ratio is three to four orders of magnitude higher than the other redox couples such as $NADPH/NADP^+$, $NADH/NAD^+$, or reduced thiore-doxin/oxidized thioredoxin, the GSH/GSSH ratio is believed to be the primary cellular determinant of the cellular redox state. Therefore, a decline in the GSH/GSSG redox status during aging may affect the overall intracellular redox environment as well as cell viability [60, 62].

Glutathione also plays a role in mitochondrial apoptosis that is regulated by BCL2 family members. Mitochondrial GSH depletion triggers mitochondrial permeabilization and activation of caspases, leading to apoptosis [60]. Furthermore, one of the earliest biochemical symptoms in Parkinson's disease is total GSH depletion [89]. Treatment with GSH-ethyl ester, which elevates GSH levels, protected neurons against dopamine loss in a rat Parkinson's disease model [90]. Interestingly, CR is known to increase the ratio of GSH/GSSG in mitochondria [62], reduce oxidative nuclear DNA damage, increase the ratio of GSH/GSSG, and decrease GSSG levels in the inner ears [67], while overexpression of CAT reduces oxidative DNA damage in the inner ear [74]. Together, these reports suggest that during aging, ROS damage to nuclear DNA, mtDNA, membranes, enzymes, and proteins in the inner ears accumulates with age, causing cochlear cell death. In contrast, CR (or starvation) reduces oxidative damage by enhancing the antioxidant defense system protecting cochlear hair cells, SGNs, and stria vascularis cells against ROS (Fig. 3).

Fig. 3 CR-mediated reduction of oxidative damage. CR reduces oxidative damage by enhancing the antioxidant defense system and protects cochlear hair cells, SGNs, and stria vascularis cells against ROS, slowing the development of hearing loss during aging

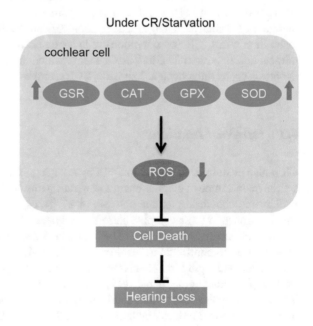

4.2   Apoptosis

ROS-induced oxidative damage is thought to promote programmed cell death or apoptosis [91], whereas CR is neuroprotective and reduces levels of apoptotic cell death. Apoptosis can occur through two major pathways. The intrinsic pathway, also known as the mitochondrial pathway, is initiated when the outer mitochondrial membrane loses its integrity, while the extrinsic pathway is initiated through ligand binding to cell surface receptors [92, 93]. In mammals, mitochondria play a major role in apoptosis, which is regulated by BCL2 family members and caspase-9. Of the BCL2 family members, the proapoptotic proteins BAX and BAK have been proposed to play a central role in promoting mitochondrial-mediated apoptosis. These BCL2 proteins promote permeabilization of the outer mitochondrial membrane, leading to cytochrome c release in the cytosol. Growing evidence suggests that apoptosis contributes to aging and age-related degenerative diseases [91, 94–96]. In an animal model of Parkinson's disease, CR lowers symptom severity and levels of apoptosis in neurons [96]. CR also reduces the levels of caspase-3 and caspase-9 in the brain of aged rats, suggesting that CR is neuroprotective [97]. Someya et al. [24] have shown that calorie-restricted mice showed a significant reduction in the number of TUNEL-positive cells and cleaved caspase-3-positive SGNs compared to age-matched controls. DNA microarray analysis also revealed that CR downregulated the expression of 24 apoptotic genes, including *Bak* (BCL2-antagonist/killer 1) and *Bim* (BCL2-like 11). Moreover, a mitochondrially targeted CAT transgene suppressed mRNA expression of *Bak* in the cochlea and reduced cochlear cell death [24]. Together, these reports suggest that the age-related accumulation of oxidative damage leads to cochlear cell death via mitochondrial apoptosis. In contrast, CR (or starvation) reduces oxidative damage by enhancing the mitochondrial antioxidant defense system, which in turn blocks mitochondrial apoptosis protecting cochlear hair cells, SGNs, and stria vascularis cells (Fig. 4).

4.3   mtDNA Mutations

CR is also postulated to protect mtDNA by reducing ROS production. Mitochondria are the main source of cellular energy, generating most cellular ATP [98]. The mammalian mitochondrial genome consists of 37 genes, encoding 13 proteins of the electron transport oxidative phosphorylation system. Because mtDNA plays essential roles in energy metabolism and cellular apoptosis, mtDNA mutations have been hypothesized to contribute to mammalian aging [94, 98]. mtDNA point mutations accumulate with aging in humans [99], and the accumulation of mtDNA mutations leads to premature aging in mice [95]. A role for mitochondrial dysfunction in AHL is supported by the observation that hearing loss is a common symptom in patients harboring inherited mtDNA mutations [100]. Specific mtDNA point mutations also contribute to mitochondrial disorders in humans such as MELAS (mitochondrial

Fig. 4 CR-mediated reduction of mitochondrial apoptosis. CR reduces oxidative damage by enhancing the antioxidant defense system, which in turn blocks mitochondrial apoptosis, protecting cochlear hair cells, SGNs, and stria vascularis cells

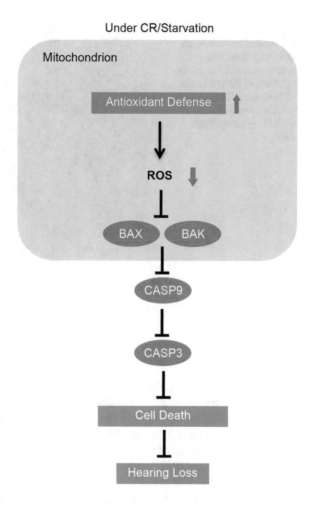

encephalomyopathy, lactic acidosis, and stroke-like episodes) and MERRF (myoclonic epilepsy and ragged red fibers), the symptoms of which include hearing loss [101–103]. More than 100 different deletions of mtDNA have been associated with mitochondrial disorders, and of these, some mtDNA deletions cause Kearns–Sayre syndrome, which involves hearing loss [103]. Moreover, several mutations in the *Polg* gene (mitochondrial DNA polymerase gamma) have been identified as a cause of human disorders such as Alpers syndrome associated with deafness [104, 105]. In mice, a knock-in mutation (D257A) that disrupts the exonuclease domain of the *Polg* gene also results in elevated levels of mitochondrial DNA point mutations and deletions, and increased levels of apoptotic cells in multiple tissues resulting in early onset AHL [95, 106]. DNA microarray analysis of cochleae from these mitochondrial DNA mutator mice revealed transcriptional alterations consistent with impairment of energy metabolism, induction of apoptosis, cytoskeletal dysfunction, and hearing dysfunction [106]. This was associated with elevated levels of apoptosis

Fig. 5 CR-mediated
reduction of mitochondrial
DNA mutations. CR
reduces mtDNA mutations
by reducing the production
of ROS, protecting
mitochondria and
preventing cochlear cell
death

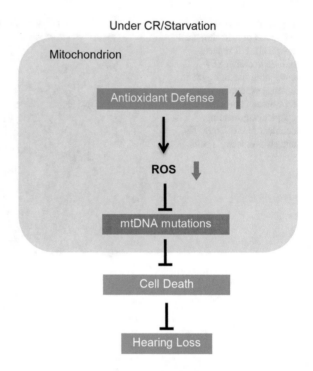

makers, including DNA fragmentation and caspase-3. These results suggest that cochlear cells are exquisitely sensitive to disturbances in energy metabolism, and that alterations in mitochondrial function with age selectively impact the cochlea, promoting apoptosis. CR is postulated to protect mtDNA by reducing the production of ROS. In support of this idea, CR reduces the levels of mtDNA deletions in the skeletal muscle [107] and liver [57] of aged rats. CR also results in reduced levels of mtDNA deletions in the auditory nerve and SV of the cochlea [36]. Collectively, these reports suggest that during aging, ROS damages mtDNA in the inner ears and mtDNA damage accumulates with age, causing mitochondrial dysfunction and cochlear cell death. In contrast, CR (or starvation) reduces mtDNA mutations by reducing the production of ROS, preventing mitochondrial dysfunction and cochlear cell death (Fig. 5).

5   Conclusion

In humans, obesity promotes a variety of age-related diseases, such as diabetes and high blood pressure [108], which are risk factors for AHL [109]. Thus, reductions in body fat and change in body composition may be key physiological changes leading to improved physical function and maintaining hearing over the course of life [110, 111]. In summary, overweight conditions likely result in increased oxidative damage in multiple tissues, including the inner ear. Overtime, this contributes to age-related cochlear cell loss via apoptosis, leading to the development of AHL. Under calorie-restricted conditions, oxidative damage to nuclear DNA, mtDNA, proteins, and lipids in the cochlea decreases and protects cochlear hair cells, SGNs, and SV cells against apoptotic cell death by slowing the development of AHL. Numerous studies have reported gender differences in human auditory perception [17, 48–54]. In general, the results of these studies show that women lose hearing more slowly than men. Thus, further studies are needed to assess the beneficial effects of CR on hearing in both males and females. Given that there are no FDA-approved medications for the treatment of AHL for older adults, empirically supported practical treatments are urgently needed to prevent hearing loss in older adults.

References

1. Fontana L, Partridge L, Longo VD (2010) Extending healthy life span-from yeast to humans. Science 328(5976):321–326. https://doi.org/10.1126/science.1172539
2. Sohal RS, Weindruch R (1996) Oxidative stress, caloric restriction, and aging. Science 273(5271):59–63. https://doi.org/10.1126/science.273.5271.59
3. Weindruch R, Sohal RS (1997) Seminars in medicine of the Beth Israel Deaconess Medical Center. Caloric intake and aging. N Engl J Med 337(14):986–994. https://doi.org/10.1056/NEJM199710023371407
4. Hindhede M (1920) The effect of food restriction during war on mortality in Copenhagen. JAMA 74(6):381–382. https://doi.org/10.1001/jama.1920.02620060015005
5. Vallejo EA (1957) Hunger diet on alternate days in the nutrition of the aged. Prensa Med Argent 44(2):119–120
6. Kagawa Y (1978) Impact of Westernization on the nutrition of Japanese: changes in physique, cancer, longevity and centenarians. Prev Med 7(2):205–217
7. Miyagi S, Iwama N, Kawabata T, Hasegawa K (2003) Longevity and diet in Okinawa, Japan: the past, present and future. Asia Pac J Public Health 15(Suppl):S3–S9. https://doi.org/10.1177/101053950301500s03
8. Anton SD, Woods AJ, Ashizawa T, Barb D, Buford TW, Carter CS, Clark DJ, Cohen RA, Corbett DB, Cruz-Almeida Y, Dotson V, Ebner N, Efron PA, Fillingim RB, Foster TC, Gundermann DM, Joseph AM, Karabetian C, Leeuwenburgh C, Manini TM, Marsiske M, Mankowski RT, Mutchie HL, Perri MG, Ranka S, Rashidi P, Sandesara B, Scarpace PJ, Sibille KT, Solberg LM, Someya S, Uphold C, Wohlgemuth S, Wu SS, Pahor M (2015) Successful aging: advancing the science of physical independence in older adults. Ageing Res Rev 24(Pt B):304–327. https://doi.org/10.1016/j.arr.2015.09.005
9. Manini TM, Buford TW, Lott DJ, Vandenborne K, Daniels MJ, Knaggs JD, Patel H, Pahor M, Perri MG, Anton SD (2014) Effect of dietary restriction and exercise on lower extremity tissue compartments in obese, older women: a pilot study. J Gerontol A Biol Sci Med Sci 69(1):101–108. https://doi.org/10.1093/gerona/gls337
10. Chen DS, Genther DJ, Betz J, Lin FR (2014) Association between hearing impairment and self-reported difficulty in physical functioning. J Am Geriatr Soc 62(5):850–856. https://doi.org/10.1111/jgs.12800
11. Gispen FE, Chen DS, Genther DJ, Lin FR (2014) Association between hearing impairment and lower levels of physical activity in older adults. J Am Geriatr Soc 62(8):1427–1433. https://doi.org/10.1111/jgs.12938

12. Han C, Ding D, Lopez MC, Manohar S, Zhang Y, Kim MJ, Park HJ, White K, Kim YH, Linser P, Tanokura M, Leeuwenburgh C, Baker HV, Salvi RJ, Someya S (2016) Effects of long-term exercise on age-related hearing loss in mice. J Neurosci 36(44):11308–11319

13. Yamasoba T, Lin FR, Someya S, Kashio A, Sakamoto T, Kondo K (2013) Current concepts in age-related hearing loss: epidemiology and mechanistic pathways. Hear Res 303:30–38. https://doi.org/10.1016/j.heares.2013.01.021

14. Gates GA, Mills JH (2005) Presbycusis. Lancet 366(9491):1111–1120. https://doi.org/10.1016/S0140-6736(05)67423-5

15. Ozmeral EJ, Eddins AC, Frisina DR, Eddins DA (2016) Large cross-sectional study of presbycusis reveals rapid progressive decline in auditory temporal acuity. Neurobiol Aging 43:72–78. https://doi.org/10.1016/j.neurobiolaging.2015.12.024

16. WHO (2019) Prevention of blindness and deafness: estimates. https://www.who.int/pbd/deafness/estimates/en/. Accessed 5 Jul 2019

17. Lin FR, Niparko JK, Ferrucci L (2011b) Hearing loss prevalence in the United States. Arch Intern Med 171(20):1851–1852. https://doi.org/10.1001/archinternmed.2011.506

18. Lin FR, Metter EJ, O'Brien RJ, Resnick SM, Zonderman AB, Ferrucci L (2011a) Hearing loss and incident dementia. Arch Neurol 68(2):214–220. https://doi.org/10.1001/archneurol.2010.362

19. Liberman MC, Kujawa SG (2017) Cochlear synaptopathy in acquired sensorineural hearing loss: Manifestations and mechanisms. Hear Res 349:138–147. https://doi.org/10.1016/j.heares.2017.01.003

20. Hudspeth AJ (1997) How hearing happens. Neuron 19(5):947–950

21. Schuknecht HF (1955) Presbycusis. Laryngoscope 65(6):402–419. https://doi.org/10.1288/00005537-195506000-00002

22. Schuknecht HF, Watanuki K, Takahashi T, Belal AA Jr, Kimura RS, Jones DD, Ota CY (1974) Atrophy of the stria vascularis, a common cause for hearing loss. Laryngoscope 84(10):1777–1821. https://doi.org/10.1288/00005537-197410000-00012

23. Gratton MA, Schulte BA (1995) Alterations in microvasculature are associated with atrophy of the stria vascularis in quiet-aged gerbils. Hear Res 82(1):44–52

24. Someya S, Yamasoba T, Weindruch R, Prolla TA, Tanokura M (2007) Caloric restriction suppresses apoptotic cell death in the mammalian cochlea and leads to prevention of presbycusis. Neurobiol Aging 28(10):1613–1622. https://doi.org/10.1016/j.neurobiolaging.2006.06.024

25. Colman RJ, Anderson RM, Johnson SC, Kastman EK, Kosmatka KJ, Beasley TM, Allison DB, Cruzen C, Simmons HA, Kemnitz JW, Weindruch R (2009) Caloric restriction delays disease onset and mortality in rhesus monkeys. Science 325(5937):201–204. https://doi.org/10.1126/science.1173635

26. Mattison JA, Roth GS, Beasley TM, Tilmont EM, Handy AM, Herbert RL, Longo DL, Allison DB, Young JE, Bryant M, Barnard D, Ward WF, Qi W, Ingram DK, de Cabo R (2012) Impact of caloric restriction on health and survival in rhesus monkeys from the NIA study. Nature 489(7415):318–321. https://doi.org/10.1038/nature11432

27. Sohal RS, Ku HH, Agarwal S, Forster MJ, Lal H (1994) Oxidative damage, mitochondrial oxidant generation and antioxidant defenses during aging and in response to food restriction in the mouse. Mech Ageing Dev 74(1–2):121–133

28. Mattson MP, Duan W, Guo Z (2003) Meal size and frequency affect neuronal plasticity and vulnerability to disease: cellular and molecular mechanisms. J Neurochem 84(3):417–431. https://doi.org/10.1046/j.1471-4159.2003.01586.x

29. Pugh TD, Klopp RG, Weindruch R (1999) Controlling caloric consumption: protocols for rodents and rhesus monkeys. Neurobiol Aging 20(2):157–165

30. Barkin RM, Todd JK, Amer J (1978) Periorbital cellulitis in children. Pediatrics 62(3):390–392

31. Keithley EM, Canto C, Zheng QY, Fischel-Ghodsian N, Johnson KR (2004) Age-related hearing loss and the ahl locus in mice. Hear Res 188(1–2):21–28. https://doi.org/10.1016/S0378-5955(03)00365-4

32. Zheng QY, Johnson KR, Erway LC (1999) Assessment of hearing in 80 inbred strains of mice by ABR threshold analyses. Hear Res 130(1–2):94–107
33. Willott JF, Erway LC, Archer JR, Harrison DE (1995) Genetics of age-related hearing loss in mice. II. Strain differences and effects of caloric restriction on cochlear pathology and evoked response thresholds. Hear Res 88(1–2):143–155
34. Sweet RJ, Price JM, Henry KR (1988) Dietary restriction and presbyacusis: periods of restriction and auditory threshold losses in the CBA/J mouse. Audiology 27(6):305–312
35. Henry KR (1986) Effects of dietary restriction on presbyacusis in the mouse. Audiology 25(6):329–337
36. Seidman MD (2000) Effects of dietary restriction and antioxidants on presbyacusis. Laryngoscope 110(5 Pt 1):727–738. https://doi.org/10.1097/00005537-200005000-00003
37. Mannström P, Ulfhake B, Kirkegaard M, Ulfendahl M (2013) Dietary restriction reduces age-related degeneration of stria vascularis in the inner ear of the rat. Exp Gerontol 48(11):1173–1179. https://doi.org/10.1016/j.exger.2013.07.004
38. Fowler CG, Torre P, Kemnitz JW (2002) Effects of caloric restriction and aging on the auditory function of rhesus monkeys (Macaca mulatta): The University of Wisconsin Study. Hear Res 169(1–2):24–35
39. Mouse Phenome Database (2007) Yuan 2: Aging study: Lifespan and survival curves for 31 inbred strains of mice. http://phenome.jax.org/pubcgi/phenome/mpdcgi?rtn=projects/details&sym=Yuan2. Accessed 5 Jul 2019
40. Bielefeld EC, Coling D, Chen GD, Li M, Tanaka C, Hu BH, Henderson D (2008) Age-related hearing loss in the Fischer 344/NHsd rat substrain. Hear Res 241(1–2):26–33. https://doi.org/10.1016/j.heares.2008.04.006
41. Keithley EM, Ryan AF, Feldman ML (1992) Cochlear degeneration in aged rats of four strains. Hear Res 59(2):171–178
42. Syka J (2010) The Fischer 344 rat as a model of presbycusis. Hear Res 264(1–2):70–78. https://doi.org/10.1016/j.heares.2009.11.003
43. Fetoni AR, Picciotti PM, Paludetti G, Troiani D (2011) Pathogenesis of presbycusis in animal models: a review. Exp Gerontol 46(6):413–425. https://doi.org/10.1016/j.exger.2010.12.003
44. Keithley EM, Feldman ML (1982) Hair cell counts in an age-graded series of rat cochleas. Hear Res 8(3):249–262
45. Alvarado JC, Fuentes-Santamaría V, Gabaldón-Ull MC, Blanco JL, Juiz JM (2014) Wistar rats: a forgotten model of age-related hearing loss. Front Aging Neurosci 6:29. https://doi.org/10.3389/fnagi.2014.00029
46. Roth GS, Mattison JA, Ottinger MA, Chachich ME, Lane MA, Ingram DK (2004) Aging in rhesus monkeys: relevance to human health interventions. Science 305(5689):1423–1426. https://doi.org/10.1126/science.1102541
47. Torre P, Fowler CG (2000) Age-related changes in auditory function of rhesus monkeys (Macaca mulatta). Hear Res 142(1–2):131–140
48. Caras ML (2013) Estrogenic modulation of auditory processing: a vertebrate comparison. Front Neuroendocrinol 34(4):285–299. https://doi.org/10.1016/j.yfrne.2013.07.006
49. Chung DY, Mason K, Gannon RP, Willson GN (1983) The ear effect as a function of age and hearing loss. J Acoust Soc Am 73(4):1277–1282. https://doi.org/10.1121/1.389276
50. Dehan CP, Jerger J (1990) Analysis of gender differences in the auditory brainstem response. Laryngoscope 100(1):18–24. https://doi.org/10.1288/00005537-199001000-00005
51. Jerger J, Johnson K (1998) Interactions of age, gender, and sensorineural hearing loss on ABR latency. Ear Hear 9(4):168–176
52. Jönsson R, Rosenhall U, Gause-Nilsson I, Steen B (1998) Auditory function in 70- and 75-year-olds of four age cohorts. A cross-sectional and time-lag study of presbyacusis. Scand Audiol 27(2):81–93
53. McFadden D (1998) Sex differences in the auditory system. Dev Neuropsychol 14:261–298. https://doi.org/10.1080/87565649809540712

54. Snihur AW, Hampson E (2011) Sex and ear differences in spontaneous and click-evoked otoacoustic emissions in young adults. Brain Cogn 77(1):40–47. https://doi.org/10.1016/j. bandc.2011.06.004

55. Torre P, Mattison JA, Fowler CG, Lane MA, Roth GS, Ingram DK (2004) Assessment of auditory function in rhesus monkeys (Macaca mulatta): effects of age and calorie restriction. Neurobiol Aging 25(7):945–954. https://doi.org/10.1016/j.neurobiolaging.2003.09.006

56. Balaban RS, Nemoto S, Finkel T (2005) Mitochondria, oxidants, and aging. Cell 120(4):483–495. https://doi.org/10.1016/j.cell.2005.02.001

57. Cassano P, Lezza AM, Leeuwenburgh C, Cantatore P, Gadaleta MN (2004) Measurement of the 4,834-bp mitochondrial DNA deletion level in aging rat liver and brain subjected or not to caloric restriction diet. Ann N Y Acad Sci 1019:269–273. https://doi.org/10.1196/annals.1297.045

58. Beckman KB, Ames BN (1998) The free radical theory of aging matures. Physiol Rev 78(2):547–581. https://doi.org/10.1152/physrev.1998.78.2.547

59. Evans P, Halliwell B (1999) Free radicals and hearing. Cause, consequence, and criteria. Ann N Y Acad Sci 884:19–40. https://doi.org/10.1111/j.1749-6632.1999.tb08633.x

60. Marí M, Morales A, Colell A, Garcia-Ruiz C, Fernandez-Checa JC (2009) Mitochondrial glutathione, a key survival antioxidant. Antioxid Redox Signal 11(11):2685-2700. https://doi.org/10.1089/ars.2009.2695

61. Beckman KB, Ames BN (1997) Oxidative decay of DNA. J Biol Chem 272(32):19633–19636. https://doi.org/10.1074/jbc.272.32.19633

62. Rebrin I, Sohal RS (2008) Pro-oxidant shift in glutathione redox state during aging. Adv Drug Deliv Rev 60(13–14):1545–1552. https://doi.org/10.1016/j.addr.2008.06.001

63. Sun J, Folk D, Bradley TJ, Tower J (2002) Induced overexpression of mitochondrial Mn-superoxide dismutase extends the life span of adult Drosophila melanogaster. Genetics 161(2):661–672

64. Runko AP, Griswold AJ, Min KT (2008) Overexpression of frataxin in the mitochondria increases resistance to oxidative stress and extends lifespan in Drosophila. FEBS Lett 582(5):715–719. https://doi.org/10.1016/j.febslet.2008.01.046

65. Schriner SE, Linford NJ, Martin GM, Treuting P, Ogburn CE, Emond M, Coskun PE, Ladiges W, Wolf N, Van Remmen H, Wallace DC, Rabinovitch PS (2005) Extension of murine life span by overexpression of catalase targeted to mitochondria. Science 308(5730):1909–1911. https://doi.org/10.1126/science.1106653

66. Liu XZ, Yan D (2007) Ageing and hearing loss. J Pathol 211(2):188–197. https://doi.org/10.1002/path.2102

67. Someya S, Prolla TA (2010) Mitochondrial oxidative damage and apoptosis in age-related hearing loss. Mech Ageing Dev 131(7–8):480–486. https://doi.org/10.1016/j.mad.2010.04.006

68. Yamasoba T, Someya S, Yamada C, Weindruch R, Prolla TA, Tanokura M (2007) Role of mitochondrial dysfunction and mitochondrial DNA mutations in age- related hearing loss. Hear Res 226(1–2):185–193. https://doi.org/10.1016/j.heares.2006.06.004

69. Jacono AA, Hu B, Kopke RD, Henderson D, Van De Water TR, Steinman HM (1998) Changes in cochlear antioxidant enzyme activity after sound conditioning and noise exposure in the chinchilla. Hear Res 117(1–2):31–38

70. McFadden SL, Ding D, Reaume AG, Flood DG, Salvi RJ (1999) Age-related cochlear hair cell loss is enhanced in mice lacking copper/zinc superoxide dismutase. Neurobiol Aging 20(1):1–8

71. Ohlemiller KK, McFadden SL, Ding DL, Lear PM, Ho YS (2000) Targeted mutation of the gene for cellular glutathione peroxidase (Gpx1) increases noise-induced hearing loss in mice. J Assoc Res Otolaryngol 1(3):243–254. https://doi.org/10.1007/s101620010043

72. Fortunato G, Marciano E, Zarrilli F, Mazzaccara C, Intrieri M, Calcagno G, Vitale DF, La Manna P, Saulino C, Marcelli V, Sacchetti L (2004) Paraoxonase and superoxide dismutase

gene polymorphisms and noise-induced hearing loss. Clin Chem 50(11):2012–2018. https://doi.org/10.1373/clinchem.2004.037788
73. Jiang H, Talaska AE, Schacht J, Sha SH (2007) Oxidative imbalance in the aging inner ear. Neurobiol Aging 28(10):1605–1612. https://doi.org/10.1016/j.neurobiolaging.2006.06.025
74. Someya S, Xu J, Kondo K, Ding D, Salvi RJ, Yamasoba T, Rabinovitch PS, Weindruch R, Leeuwenburgh C, Tanokura M, Prolla TA (2009) Age-related hearing loss in C57BL/6J mice is mediated by Bak-dependent mitochondrial apoptosis. Proc Natl Acad Sci U S A 106(46):19432–19437. https://doi.org/10.1073/pnas.0908786106
75. Van Eyken E, Van Camp G, Fransen E, Topsakal V, Hendrickx JJ, Demeester K, Van de Heyning P, Maki-Torkko E, Hannula S, Sorri M, Jensen M, Parving A, Bille M, Baur M, Pfister M, Bonaconsa A, Mazzoli M, Orzan E, Espeso A, Stephens D, Verbruggen K, Huyghe J, Dhooge I, Huygen P, Kremer H, Cremers CW, Kunst S, Manninen M, Pyykko I, Lacava A, Steffens M, Wienker TF, Van Laer L (2007) Contribution of the N-acetyltransferase 2 polymorphism NAT2*6A to age- related hearing impairment. J Med Genet 44(9):570–578. https://doi.org/10.1136/jmg.2007.049205
76. Bared A, Ouyang X, Angeli S, Du LL, Hoang K, Yan D, Liu XZ (2010) Antioxidant enzymes, presbycusis, and ethnic variability. Otolaryngol Head Neck Surg 143(2):263–268. https://doi.org/10.1016/j.otohns.2010.03.024
77. Banaclocha MM (2001) Therapeutic potential of N-acetylcysteine in age-related mitochondrial neurodegenerative diseases. Med Hypotheses 56(4):472–477. https://doi.org/10.1054/mehy.2000.1194
78. Palaniappan AR, Dai A (2007) Mitochondrial ageing and the beneficial role of alpha-lipoic acid. Neurochem Res 32(9):1552–1558. https://doi.org/10.1007/s11064-007-9355-4
79. Hagen TM, Moreau R, Suh JH, Visioli F (2002) Mitochondrial decay in the aging rat heart: evidence for improvement by dietary supplementation with acetyl-L-carnitine and/or lipoic acid. Ann N Y Acad Sci 959:491–507. https://doi.org/10.1111/j.1749-6632.2002.tb02119.x
80. Liu J, Killilea DW, Ames BN (2002) Age-associated mitochondrial oxidative decay: improvement of carnitine acetyltransferase substrate-binding affinity and activity in brain by feeding old rats acetyl-L- carnitine and/or R-alpha -lipoic acid. Proc Natl Acad Sci U S A 99(4):1876–1881. https://doi.org/10.1073/pnas.261709098
81. Derin A, Agirdir B, Derin N, Dinc O, Guney K, Ozcaglar H, Kilincarslan S (2004) The effects of L-carnitine on presbyacusis in the rat model. Clin Otolaryngol Allied Sci 29(3):238–241. https://doi.org/10.1111/j.1365-2273.2004.00790.x
82. Seidman MD, Khan MJ, Bai U, Shirwany N, Quirk WS (2000) Biologic activity of mitochondrial metabolites on aging and age-related hearing loss. Am J Otol 21(2):161–167
83. Ahn JH, Kang HH, Kim TY, Shin JE, Chung JW (2008) Lipoic acid rescues DBA mice from early-onset age-related hearing impairment. Neuroreport 19(13):1265–1269. https://doi.org/10.1097/WNR.0b013e328308b338
84. Ding D, Jiang H, Chen GD, Longo-Guess C, Muthaiah VP, Tian C, Sheppard A, Salvi R, Johnson KR (2014) N-acetyl-cysteine prevents age-related hearing loss and the progressive loss of inner hair cells in γ-glutamyl transferase 1 deficient mice. Aging (Albany NY) 8(4):730–750. https://doi.org/10.18632/aging
85. Wu HP, Hsu CJ, Cheng TJ, Guo YL (2010) N-acetylcysteine attenuates noise-induced permanent hearing loss in diabetic rats. Hear Res 267(1–2):71–77. https://doi.org/10.1016/j.heares.2010.03.082
86. Ohinata Y, Yamasoba T, Schacht J, Miller JM (2000) Glutathione limits noise-induced hearing loss. Hear Res 146(1–2):28–34
87. Sohal RS, Forster MJ (2007) Coenzyme Q, oxidative stress and aging. Mitochondrion 7(Suppl):S103–S111. https://doi.org/10.1016/j.mito.2007.03.006
88. Guastini L, Mora R, Dellepiane M, Santomauro V, Giorgio M, Salami A (2011) Water-soluble coenzyme Q10 formulation in presbycusis: long-term effects. Acta Otolaryngol 131(5):512–517. https://doi.org/10.3109/00016489.2010.539261

89. Perry TL, Yong VW (1986) Idiopathic Parkinson's disease, progressive supranuclear palsy and glutathione metabolism in the substantia nigra of patients. Neurosci Lett 67(3):269–274. https://doi.org/10.1016/0304-3940(86)90320-4

90. Zeevalk GD, Bernard LP, Song C, Gluck M, Ehrhart J (2005) Mitochondrial inhibition and oxidative stress: reciprocating players in neurodegeneration. Antioxid Redox Signal 7(9–10):1117–1139. https://doi.org/10.1089/ars.2005.7.1117

91. Dirks AJ, Hofer T, Marzetti E, Pahor M, Leeuwenburgh C (2006) Mitochondrial DNA mutations, energy metabolism and apoptosis in aging muscle. Ageing Res Rev 5(2):179–195. https://doi.org/10.1016/j.arr.2006.03.002

92. Lindsten T, Ross AJ, King A, Zong WX, Rathmell JC, Shiels HA, Ulrich E, Waymire KG, Mahar P, Frauwirth K, Chen Y, Wei M, Eng VM, Adelman DM, Simon MC, Ma A, Golden JA, Evan G, Korsmeyer SJ, MacGregor GR, Thompson CB (2000) The combined functions of proapoptotic Bcl-2 family members bak and bax are essential for normal development of multiple tissues. Mol Cell 6(6):1389–1399

93. Youle RJ, Strasser A (2008) The BCL-2 protein family: opposing activities that mediate cell death. Nat Rev Mol Cell Biol 9(1):47–59. https://doi.org/10.1038/nrm2308

94. Kujoth GC, Bradshaw PC, Haroon S, Prolla TA (2007) The role of mitochondrial DNA mutations in mammalian aging. PLoS Genet 3(2):e24. https://doi.org/10.1371/journal.pgen.0030024

95. Kujoth GC, Hiona A, Pugh TD, Someya S, Panzer K, Wohlgemuth SE, Hofer T, Seo AY, Sullivan R, Jobling WA, Morrow JD, Van Remmen H, Sedivy JM, Yamasoba T, Tanokura M, Weindruch R, Leeuwenburgh C, Prolla TA (2005) Mitochondrial DNA mutations, oxidative stress, and apoptosis in mammalian aging. Science 309(5733):481–484. https://doi.org/10.1126/science.1112125

96. Mattson MP (2000) Apoptosis in neurodegenerative disorders. Nat Rev Mol Cell Biol 1(2):120–129. https://doi.org/10.1038/35040009

97. Shelke RR, Leeuwenburgh C (2003) Lifelong caloric restriction increases expression of apoptosis repressor with a caspase recruitment domain (ARC) in the brain. FASEB J 17(3):494–496. https://doi.org/10.1096/fj.02-0803fje

98. Wallace DC, Shoffner JM, Trounce I, Brown MD, Ballinger SW, Corral-Debrinski M, Horton T, Jun AS, Lott MT (1995) Mitochondrial DNA mutations in human degenerative diseases and aging. Biochim Biophys Acta 1271(1):141–151. https://doi.org/10.1016/0925-4439(95)00021-u

99. Michikawa Y, Mazzucchelli F, Bresolin N, Scarlato G, Attardi G (1999) Aging-dependent large accumulation of point mutations in the human mtDNA control region for replication. Science 286(5440):774–779. https://doi.org/10.1126/science.286.5440.774

100. Chinnery PF, Elliott C, Green GR, Rees A, Coulthard A, Turnbull DM, Griffiths TD (2000) The spectrum of hearing loss due to mitochondrial DNA defects. Brain 123(Pt 1):82–92. https://doi.org/10.1093/brain/123.1.82

101. Fischel-Ghodsian N (2003) Mitochondrial deafness. Ear Hear 24(4):303–313. https://doi.org/10.1097/01.aud.0000079802.82344.b5

102. Kokotas H, Petersen MB, Willems PJ (2007) Mitochondrial deafness. Clin Genet 71(5):379–391. https://doi.org/10.1111/j.1399-0004.2007.00800.x

103. Pickles JO (2004) Mutation in mitochondrial DNA as a cause of presbyacusis. Audiol Neurootol 9(1):23–33. https://doi.org/10.1159/000074184

104. Mancuso M, Filosto M, Bellan M, Liguori R, Montagna P, Baruzzi A, DiMauro S, Carelli V (2004) POLG mutations causing ophthalmoplegia, sensorimotor polyneuropathy, ataxia, and deafness. Neurology 62(2):316–318. https://doi.org/10.1212/wnl.62.2.316

105. Nguyen KV, Ostergaard E, Ravn SH, Balslev T, Danielsen ER, Vardag A, McKiernan PJ, Gray G, Naviaux RK (2005) POLG mutations in Alpers syndrome. Neurology 65(9):1493–1495. https://doi.org/10.1212/01.wnl.0000182814.55361.70

106. Someya S, Yamasoba T, Kujoth GC, Pugh TD, Weindruch R, Tanokura M, Prolla TA (2008) The role of mtDNA mutations in the pathogenesis of age-related hearing loss in mice

carrying a mutator DNA polymerase gamma. Neurobiol Aging 29(7):1080–1092. https://doi.org/10.1016/j.neurobiolaging.2007.01.014

107. Aspnes LE, Lee CM, Weindruch R, Chung SS, Roecker EB, Aiken JM (1997) Caloric restriction reduces fiber loss and mitochondrial abnormalities in aged rat muscle. FASEB J 11(7):573–581. https://doi.org/10.1096/fasebj.11.7.9212081

108. Paeratakul S, Lovejoy JC, Ryan DH, Bray GA (2002) The relation of gender, race and socio-economic status to obesity and obesity comorbidities in a sample of US adults. Int J Obes Relat Metab Disord 26(9):1205–1210. https://doi.org/10.1038/sj.ijo.0802026

109. Vaughan N, James K, McDermott D, Griest S, Fausti S (2006) A 5-year prospective study of diabetes and hearing loss in a veteran population. Otol Neurotol 27(1):37–43

110. Dutta D, Calvani R, Bernabei R, Leeuwenburgh C, Marzetti E (2012) Contribution of impaired mitochondrial autophagy to cardiac aging: mechanisms and therapeutic opportunities. Circ Res 110(8):1125–1138. https://doi.org/10.1161/circresaha.111.246108

111. Ivanova DG, Yankova TM (2013) The free radical theory of aging in search of a strategy for increasing life span. Folia Med (Plovdiv) 55(1):33–41

Noise-Induced Hearing Loss and Drug Therapy: Basic and Translational Science

Celia Escabi, Monica Trevino, Eric Bielefeld, and Edward Lobarinas

1 Introduction

Noise is among the most prevalent causes of sensorineural hearing loss in both the adolescent and adult population. In industrialized societies, noise sources are common in the workplace, transportation, recreation, and in home settings. Sixteen percent of disabling hearing loss in the world can be attributed to occupational noise exposure [1]. Occupational noise standards in the United States were predicated upon the notion that workers were not getting significant noise doses outside the workplace. However, avoiding noise in nonoccupational settings has become difficult or impossible for many workers around the world. Those nonoccupational noise sources are likely contributing to the prevalence of noise-induced hearing loss (NIHL) (Table 1). Survey data from 2005 to 2006 indicated that significant NIHL was present in 16.8% of Americans between 12 and 19 years of age [2].

The best method of preventing NIHL is via acoustic attenuation from personal hearing protection devices (PHPDs). The sound attenuation from PHPDs is often effective at reducing the noise doses at the cochlea down below the level that can cause injury, but there are many circumstances in which the noise-exposed individual cannot or will not use a PHPD. An alternative to PHPD use is rendering the ear less susceptible to noise injury via pharmaceutical intervention. The goal of these otoprotective treatments is to interrupt noise-induced biological mechanisms that normally lead to injury and death of cells in the organ of Corti, spiral ganglion, and/or the lateral wall. For unanticipated noise exposures, the treatment must occur

C. Escabi · M. Trevino · E. Lobarinas (✉)
School of Behavioral and Brain Sciences, The University of Texas at Dallas, Callier Center for Communication Disorders, Dallas, TX, USA
e-mail: edward.lobarinas@utdallas.edu

E. Bielefeld
Department of Speech and Hearing Science, The Ohio State University, Columbus, OH, USA

© Springer Nature Switzerland AG 2020
S. Pucheu et al. (eds.), *New Therapies to Prevent or Cure Auditory Disorders*,
https://doi.org/10.1007/978-3-030-40413-0_2

Table 1 List of abbreviations used in the chapter

	Abbreviation		Abbreviation
Auditory brainstem response	ABR	Nicotinamide adenine dinucleotide phosphate	NADPH
Auditory nerve fibers	ANFs	Noise-induced hearing loss	NIHL
Action potential	AP	N-Methyl-D-aspartic acid	NMDA
Auditory steady state response	ASSR	Outer hair cell	OHC
Adenosine triphosphate	ATP	Personal hearing protection devices	PHPDs
Compound action potential	CAP	Protein-tyrosine kinase	PTK
Decibel sound pressure level	dB SPL	Permanent threshold shift	PTS
Dehydroepiandrosterone	DHEAS	R-Phenylisopropyladenosine	R-PIA
Deoxyribonucleic acid	DNA	Reactive oxygen species	ROS
Distortion product otoacoustic emissions	DPOAE	Spiral ganglion neurons	SGNs
Extended high frequency	EHF	Signal-to-noise ratio	SNR
Extended high-frequency audiometry	EHFA	Superoxide dismutase	SOD
Inner hair cell	IHC	Summating potential	SP
c-Jun N-terminal kinase	JNK	Sound pressure level	SPL
Lifetime noise exposure	LNE	Spontaneous rate	SR
n-Acetyl, L-cysteine	NAC	Temporary threshold shift	TTS

after the noise exposure has already begun. These otoprotection treatments are referred to as "rescue," because they attempt to mitigate damage that has already begun to manifest.

Pharmaceutical otoprotection first requires finding a compound that effectively protects the cochlea against injury. As discussed in Sect. 2, identifying candidate compounds requires a good understanding of the mechanisms of noise-induced cochlear injury. One of the factors limiting the discovery of effective otoprotective compounds is the lack of available high-throughput screening tools. The zebrafish lateral line [3–5], in vitro cultures of the organ of Corti [6, 7] and in vitro cultures of the spiral ganglion [8,9] can be used to screen for potential ototoxic compounds or otoprotective compounds that reduce ototoxicity. However, NIHL is purely an in vivo phenomenon that requires animal models using different types of noise. This makes investigation into pharmaceutical otoprotection costly and time-consuming.

In addition to discovering effective compounds, there are other critical variables including drug dose levels, timing of the doses, and route of delivery that must be considered when developing a drug for otoprotection. Because the organ of Corti is behind the restrictive blood–labyrinth barrier, it is challenging to deliver compounds systemically that achieve entry into the cochlea at sufficient levels to induce otoprotection. Poor cochlear bioavailability necessitates higher doses, which can lead to systemic side effects that limit the utility of the compounds. Alternatively, local dosing directly into the cochlea offers considerable potential. With local drug delivery, the middle ear space is accessed through the tympanic membrane or the auditory bulla. In some cases, the compound is delivered into the cochlea itself or applied to the middle ear where it diffuses through the round window membrane. These approaches have clear advantages, such as largely eliminating systemic toxicity and allowing lower dose levels to be used. The local dosing approach is often used with animal models to identify compounds that can be otoprotective against noise. The challenge with local cochlear drug delivery is the invasive nature of the procedures and the difficulties they present in terms of regular, long-term drug delivery.

Section 6 of this chapter will review a variety of otoprotection strategies, grouped by the class of compounds used and the cochlear injury pathways that these attempt to intercede. Before those compounds are discussed, an overview of the patterns, mechanisms, and consequences of noise-induced cochlear injury is required.

2   Cochlear Pathology

Noise exposures can damage many different cochlear structures, including the organ of Corti, neuronal synapses, spiral ligament, and stria vascularis [10–12]. A major consequence of noise exposure is outer hair cell (OHC) death, which often produces permanent threshold shift (PTS). With severe and persistent NIHL, inner hair cell (IHC) death can also occur, and in some animal species, it is better correlated with PTS [13, 14] than OHC loss is. Further, noise can cause temporary and permanent injury to the spiral ligament fibrocytes [12] and stria vascularis [15], which can disrupt the maintenance of the endocochlear potential. Noise can also cause injury to the afferent synapses that contact the IHCs and the dendrites of the spiral ganglion neurons, a phenomenon that is discussed in more detail in Sect. 3.

Cochlear injuries can be broadly classified as metabolic or mechanical. Metabolic injuries reflect a disruption of homeostasis from internal or external stressors to the cell. Noise-induced metabolic injuries can occur due to oxidative stress [16–19], excitotoxicity [20], or apoptosis pathways [21–24], though this is not an exhaustive list of traumatic events that can lead to NIHL.

2.1   Oxidative Stress

Oxidative stress, in the inner ear, can occur because of an imbalance between reactive oxygen species (ROS) and antioxidants within a cell population (reviewed in Henderson et al. [10] and Le Prell et al. [25, 26]). It often results from either an overproduction of ROS and/or a depletion of antioxidants. Excessive ROS can break down cell membranes through lipid peroxidation, alter proteins or deoxyribonucleic acid (DNA), and act as a putative trigger for apoptotic cell death. Increased ROS formation has been documented following noise exposure in multiple animal studies [16–19]. Although there is currently no single definitive explanation, several mechanisms could account for increased ROS formation, including ischemia/reperfusion from cochlear blood flow changes, excitotoxicity, nitric oxide activity [27, 28], increases in nicotinamide adenine dinucleotide phosphate (NADPH) oxidase activity [29–31], and mitochondrial dysfunction [32].

2.2 Cochlear Blood Flow Changes

The relationship between noise and cochlear blood flow is complex. Noise is a system-wide stressor that triggers increased activation of the sympathetic nervous system [33], leading to increased cardiac output and blood pressure. Increased system-wide blood flow is typically accompanied by increased blood flow in the cochlea. However, these increases can be countered by vasoconstriction of the arteries that supply the cochlea. Vasoconstriction occurs in response to stimulation of the local sympathetic branches that supply the cochlear arteries, specifically the stellate ganglion and superior cervical ganglion [34, 35]. These nerve branches have been implicated in the cochlea's susceptibility to NIHL [36–40] via their effects on cochlear blood flow and influence on the afferent spiral ganglion neurons (SGNs). Overall, noise has been shown to lower cochlear blood flow in numerous animal models [41–43]. Whereas it is not clear that blood flow changes are causing oxidative stress, it is reasonable to expect that these are at least modulating oxidative stress and may indeed exacerbate it.

2.3 Apoptotic Cell Death

Apoptosis is a programmed sequence of intracellular signaling events that culminates in cell death [44]. The outcome is cell death without loss of integrity of the cell membrane, thus reducing the impact of cell death on the neighboring cells and preventing a strong inflammatory response. Apoptosis has been identified as a major pathway in noise-induced cochlear injury [21, 23]. Apoptosis is an active cell death sequence that requires adenosine triphosphate (ATP) consumption, and the direction of a cell toward apoptosis rather than necrosis (loss of cellular membrane integrity leading to cell lysis and death) is dictated by the availability of ATP. Insufficient ATP forces the cell into necrosis [45]. When cochlear apoptosis occurs, intracellular signaling begins at the level of the cell membrane [46]; nucleus; or mitochondria [24], prompted by internal or external stressors, including ROS. Caspases, a family of protease enzymes that play essential roles in programmed cell death and inflammation, are signaling molecules that trigger cochlear apoptosis, which can lead to the breakdown of the cells in the organ of Corti after noise exposure [22, 24, 47]. In cases of short-duration noise exposures, there may be an expansion of the hair cell lesion shortly after the exposure. The expansion of the lesion predominantly occurs through apoptosis [48, 49].

2.4 Mechanical Damage and Stereocilia Injury

Noise-induced mechanical injuries are the result of physical damage to the delicate and precise architecture of the organ of Corti due to unusually large vertical displacements of the basilar membrane that occur in response to high sound pressure level (SPL) exposure. Mechanical injuries include broken and fused stereocilia [50], separation of hair cells from supporting cells [10], development of holes in the reticular lamina [51], and injury or loss of pillar cells [52]. For extremely high peak exposure levels, the mechanical damage can cause the basilar membrane to detach from the osseous spiral lamina [53]. This detachment disrupts the normal mechanical movement of the basilar membrane driven. Stereocilia injury can disrupt the mechanoelectric transduction process that regulates the depolarization and hyperpolarization of cochlear hair cells. Separation of the cells from their supporting cells can stress cell junctions and the extracellular matrix, which in turn can trigger apoptosis (see Sect. 2.3). In addition, mechanical damage from high sound exposure levels can produce holes in the reticular lamina that lead to a mixing of high-potassium endolymph above the lamina with the cortilymph below. This mixing can disrupt the flow of potassium ions into and out of the hair cells and result in potassium-induced toxicity. Collapse of other supporting structures such as the pillar cells or the tearing of the basilar membrane can also disrupt the precise organization of the organ of Corti and subsequently result in NIHL.

3 Noise-Induced Deafferentation

Noise exposure can also produce less obvious signs of cochlear damage by inducing deafferentation of the IHCs at the level of the SGNs and their afferent dendritic synapses. This can occur even in the absence of hair cell loss. In the acute period after a noise exposure, a reduction in type I afferent ribbon synapses has been found at the base of the IHCs where connections are made with afferent auditory nerve fibers [54, 55]. With respect to hearing sensitivity, deafferentation may manifest only as temporary threshold shift (TTS) [20, 56, 57] with no subsequent PTS. Several animal studies have conclusively shown noise-induced loss of IHC afferent synapses in the absence of either OHC or IHC loss [54, 55, 58]. The synaptopathic loss has been suggested to be an early marker of NIHL, as well as a correlate of age-related neural degeneration of the inner ear in both animals and humans. Whereas the animal data have been directly measured, human data have been derived from archival human temporal bone specimens [59, 60]. If this synaptopathic loss indeed occurs in humans as a function of noise or age, then the following questions develop: what are its consequences, how is it measured, and when does it begin?

Initial studies showing synaptopathic loss were performed on mice [55, 58]. These studies revealed that noise exposures that produced no evidence of hair cell loss or changes in hearing thresholds induced a permanent reduction in afferent IHC synapses. Similar results of synaptic loss were also found in aging studies on mice [61–63]. This major finding suggested that noise and aging can produce a form of preclinical sensorineural hearing loss that precedes hair cell loss. The SPLs required to produce synaptopathic loss in mice also generated a robust TTS on the order of 30 dB or more at 24 h postexposure, but do not produce PTS. Although auditory brainstem response (ABR) thresholds recover by 1–2 weeks, the ABR Wave I amplitude input–output function is permanently reduced after the synaptopathic noise exposure. This ABR amplitude reduction has been attributed to loss of high-threshold, low-spontaneous rate (SR) auditory nerve fibers (ANFs). Thus, the synaptopathic noise exposures preferentially damage input to the subset of ANFs activated by high-intensity acoustic input. In contrast, high-SR ANFs, activated by low-intensity sound, remain unaffected. Because of their long latencies and poor synchrony, it is unclear to what degree low-SR ANFs contribute to reduced ABR Wave I amplitudes at suprathreshold presentation levels. Alternatively, reduced Wave I amplitudes could reflect damage to high- or medium-SR ANFs, which have response latencies similar to Wave I [64]. OHC electromotility, the main driver of cochlear nonlinearity essential for low thresholds, also remains unaffected with these synaptopathic noise exposures. Synaptopathic noise exposures have not been shown to have long-term effects on distortion product otoacoustic emissions (DPOAE), a physiological measure sensitive to OHC function. Thus, noise exposures that produce robust TTS in mice can permanently damage synapses, leading to a permanent reduction in ABR amplitude Wave I amplitude with no effect on thresholds, OAEs, or OHCs.

3.1   Consequences of Synaptopathy

Given a lack of PTS, it stands to reason that these noise exposures would have been dismissed as not producing significant hearing impairment or loss. Is it possible that noise exposures previously deemed safe may actually be hazardous? If so, how intense is too intense? The levels of noise used in the mouse studies were on the order of 100–106 dB SPL for 2 h using an octave band noise (8–16 kHz). Though this is certainly high intensity, it is well within the range of high-intensity recreational activities for humans (music concerts, loud sporting events, discotheques, etc.). Subsequent studies have shown similar synaptopathic loss in other species, including guinea pigs [65], gerbils [66], rats [67], chinchillas [68], and subhuman primates [69]. The anatomical consequences have been remarkably similar across the species.

The issue of the perceptual consequences of synaptopathy has been discussed considerably, and synaptopathic loss has been speculated to be the root of problems related to poorer hearing in noise, tinnitus, and even hyperacusis. The hypothesis related to poorer hearing in noise stems from the loss of high-threshold, low-SR fibers. Because these fibers are activated at higher SPLs, and hearing in noise occurs in high-intensity backgrounds, any reduction in activity of these fibers should reduce functional ability in noise. In animals, one of the closest comparative studies aimed at evaluating the functional effects of synaptopathic noise exposures was performed on rats trained to listen to narrow band stimuli centered at 8, 12, 16, 20, and 24 kHz in the presence of competing broadband noise at 40, 20, 15, or 10 dB signal-to-noise ratios (SNRs). The synaptopathic noise exposures (109 dB SPL, 8–16 kHz octave band noise, 2 h) produced a robust 30–40 dB TTS at 24 h and a permanent reduction in ABR Wave I amplitude at 16, 24, and 32 kHz [70]. The reduction in ABR amplitude was correlated with lower performance in background noise at the same three frequencies. It is important to note, however, that these differences were only found at the poorer 20 dB SNR. This condition was challenging, as the task required the detection of a 50-ms narrow band noise in the presence of a competing broad band noise. Thus, although functional deficits were seen, these only occurred after large TTS and only under the most difficult listening conditioning, effects that would not be obvious under routine testing. A separate study looked at the possibility that synaptopathy could be correlated with the emergence of tinnitus. In Mongolian gerbils exposed to noise, the subset of animals that developed behavioral evidence of tinnitus had significantly more synaptopathy than animals without tinnitus, regardless of the degree of hearing loss [71]. Although the data are preliminary, the authors suggested that the pattern of synaptopathic loss could be an underlying contributor to the development of tinnitus. In contrast, work in mice has shown that animals with ABR measures consistent with noise-induced synaptopathy did not appear to exhibit any behaviors consistent with tinnitus or hyperacusis [72]. Thus far, across rodents, it appears that (1) synaptopathic damage can be reliably induced by noise, (2) reductions in ABR Wave I amplitudes are correlated with synaptopathic loss, but (3) the speculated functional deficits are less clear than the anatomical and physiological findings.

4 Auditory Threshold Shift as a Measure of Noise-Induced Hearing Loss

Noise exposure has a measurable impact on overall hearing sensitivity. Early studies looking at the effects of noise exposure on hearing sensitivity defined the effects of TTSs as audiometric threshold shifts from baseline levels that occur immediately after noise exposure, and resolve to baseline within a designated time frame postexposure. When TTS does not resolve, PTSs are sustained. The extent of TTS and PTS is heightened by factors such as the duration of the noise exposure and the type of noise used, with longer-duration, higher-intensity, and more dynamic energy distribution producing more damage.

Though chinchilla is slightly more susceptible to NIHL than humans [73], early studies using this animal model illustrate the relationship between TTS and PTS [74–77]. Two of these studies used a 4-kHz narrow band noise at different intensity levels (80, 86, 92, and 98 dB SPL; 57, 65, 72, 80, 86, and 92 dB SPL) and measured threshold shift at intervals from 4 min to 90days postexposure [76, 77]. The degree of TTS and PTS increased as the stimulus level increased, with the highest stimulus levels producing the most TTS and PTS. This relationship has also been observed in mice exposed to octave band noise (8–16 kHz) for 2 h at various levels (94, 100, 106, 112, and 116 dB SPL) [12]. The animals exposed to the lowest level (94 dB SPL) experienced a 50-dB TTS that completely recovered, whereas the animals exposed to 116 dB SPL experienced high level of PTS. In gerbils, TTS was measurable just 30 min after exposure to two-octave band noise at 100, 110, and 120 dB SPL; animals that experienced higher TTS also experienced higher PTS [78]. These findings are consistent across species, and show that the final TTS and PTS results vary as a function of exposure duration and intensity. The time course data have been used to generate predictive models, whereby TTS at 24 h postexposure was shown to be a relatively reliable predictor of PTS [74].

The effect of noise exposure on the audiogram has distinct features. Noise exposure typically damages hearing sensitivity at frequencies above the exposure [79]. The hallmark of a human noise-exposed ear is a "noise notch" hearing loss that occurs from 3 to 6 kHz. This characteristic pattern is believed to be the result of the average resonance of the human ear canal, which is around 3 kHz [80]. Similar effects have been shown in animal studies using chinchillas [81] and mice [12], albeit at higher frequencies in the case of the mouse. However, it is important to note that not all individuals develop characteristic notches following noise trauma. For instance, researchers analyzing data from the Nord Trøndelag Health Study examined the audiograms of nearly 50,000 individuals and found that the prevalence of notches from 3 to 6 kHz in occupationally noise-exposed individuals was 60–70%, and was 50–60% in participants without occupational noise exposure [82]. These data suggest that the "noise notch" pattern of hearing loss is prevalent, but not exclusive to noise-exposed individuals. Nonetheless, when hearing loss is evaluated in conjunction with a history of noise exposure, the "noise notch" helps point to the possible etiology of an individual's hearing loss.

4.1   Extended High-Frequency Audiometry

Although human frequency sensitivity extends to 20 kHz, pure tone thresholds in quiet environment are assessed clinically from frequencies 0.25 to 8 kHz, as this frequency range is essential for speech intelligibility. Extended high-frequency audiometry (EHFA), or assessment of thresholds from 9 to 20 kHz, has been investigated as a potential early marker for NIHL and is often used for early detection of ototoxicity in cancer patients undergoing chemotherapy. Animal studies suggest that sensory cells at the basal end of the cochlea are more susceptible to the damaging effects of noise [83]; in humans, these effects are exacerbated by external ear amplification in the 3–6 kHz range. One study compared cement factory workers regularly exposed to >85 dB SPL noise to their unexposed colleagues [84]. The study found that the noise-exposed group had higher EHF thresholds than the nonexposed group, and these threshold differences were most significant across younger age groups (21 to 30 years old, 31–40, 41–50) but were less prominent among the oldest group (51 to 60 years old), suggesting that elevation of EHF thresholds may be an early marker of occupational noise exposure damage in younger workers. On average, noise-exposed employees in the youngest group exhibited 6–20 dB higher thresholds than their unexposed age-matched peers. These results were supported by a subsequent study that sought to apply EHFA as an early screening method for NIHL in 151 at-risk workers [85]. The researchers found a similar interaction between history of noise exposure, age, and EHFA thresholds and determined that EFHA more effectively revealed significant threshold changes in employees who had worked less than 10 years compared to conventional audiometry [85]. These studies and others [86, 87] indicate that NIHL initially impacts EHFs, and that EHFA should be considered for NIHL-monitoring protocols.

5 Effects of Noise on Suprathreshold Measures of Hearing

A variety of audiological assessments have been used to evaluate potential otoprotective agents against NIHL. Evidence for efficacy in animal models of NIHL is typically based on the prevention of ABR threshold shifts or DPOAE amplitude changes [88]. In contrast, behavioral assessments with pure-tone air conduction audiometry are considered the accepted norm in human clinical trials, which rarely depend on ABR or DPOAEs as the primary outcome measure [89]. However, there is an ongoing interest in identifying other tests that may provide earlier indication of noise damage. Stimuli presented well above threshold may be more sensitive in detecting early noise-induced changes (deafferentation) relative to the behavioral audiogram, especially in individuals whose thresholds are within normal limits. A number of metrics to assess synaptopathy or "hidden hearing loss" have been proposed, including the reductions observed in ABR Wave I amplitude.

One of the first studies to look at associations between noise exposure history and ABR Wave I amplitude reported that ABR Wave I amplitude decreased as a function of increasing noise exposure history in otherwise normal-hearing individuals [90]. However, the initial study failed to accurately account for sex differences among participants. A follow-up analysis revealed no significant differences between ABR amplitudes as a function of noise exposure history [91]. Two subsequent studies were in agreement with the reanalyzed data; no significant correlations were found between ABR Wave I amplitudes and noise exposure history [92, 93].

Other measures have attempted to look at the ABR Wave I-to-Wave V ratio or the electrocochleography summating potential (SP) to action potential (AP) (SP/AP) ratio. Both have been proposed as stronger correlates of synaptopathy than ABR Wave I amplitude alone. Initial studies have suggested that these measures may correlate with both history of noise exposure and hearing-in-noise measures. In a study of college-aged students with varied noise history, participants were separated into a high-risk and low-risk group based on responses to a questionnaire that included self-perceived hearing ability, history of noise exposure, and use of hearing protection. Given that previous studies had shown the effects of sex on ABR amplitudes, ABR Wave I was normalized to the AP and compared to SP via the use of TipTrode electrodes [94]. The data were carefully analyzed, and sex differences were taken into account, particularly because the high-risk group included significantly more male participants. The overall findings showed that (1) the high-risk group had elevated high-frequency audiometric thresholds (8–16 kHz), (2) there was a significant difference in the SP/AP ratio, and (3) the high-risk group had poorer hearing-in-noise performance. Interestingly, the data showed that, though there was no significant difference in the AP, the SP amplitude was increased in the high-risk group. Thus, it appears that the SP/AP ratio may have some clinical utility in identifying early markers of NIHL, but more work is needed to understand the potential mechanisms underlying the differences and to determine if these differences are still present in the absence of high-frequency hearing loss.

A separate study [95] evaluated the effect of lifetime noise exposure (LNE) in adults (29–55 years of age) on speech-in-noise intelligibility, ABR amplitude, and Wave I-to-Wave V ratios. The findings showed a moderate relationship between LNE and reduced ABR amplitude, but did not find significant correlations between LNE and tinnitus, central gain activation, or speech-in-noise performance. Furthermore, the data were characterized as having a high degree of variability and were complicated by the effects of central gain activation as well as attention in speech-in-noise tasks. The authors concluded that, although synaptopathy could play a role in the results, it is more likely to be "at most" one of the several factors to account for differences among participants [95]. Bramhall et al. [96] investigated the use of suprathreshold ABR Wave I amplitude change and speech-in-noise performance. The study found that veterans with known high levels of noise exposure and nonveteran firearm users had reduced ABR Wave I amplitudes and greater speech-in-noise deficits compared to the lower noise-exposed group (consisting of veterans and nonveterans without a history of firearm use). The high- and low-noise groups had normal thresholds (up to 8 kHz) and similar OAEs. However, the noise-exposed veteran group tended to have elevated thresholds at 2, 3, and 4 kHz (an average of 7.3-dB HL) compared to nonveterans, suggesting sensitivity at those frequencies contributed to the differences observed between groups [96]. A study among young adult normal hearing listeners ($n = 126$) found no significant relationship between LNE and ABR Wave I amplitude measures [97], and these findings were confirmed when replicated [98]. Overall, although ABR Wave I amplitude measures provide an objective assessment, more data are needed regarding the repeatability of ABR Wave I amplitudes in humans to assess their ability to identify subclinical hearing loss. There are several problems with ABR testing in human subjects that introduce variability across individuals and within test–retest reliability. The use of Wave I/Wave V amplitude ratio could to some extent control for electrode placement differences compared to relying on the absolute measure of Wave I amplitude [99]. Other studies advocate the use of ABR Wave V latency measures instead of Wave I amplitudes, as this may be more easily measured in humans [100].

Behavioral assessments in animal models have also been conducted and reliably resemble the animals' ABR thresholds, but the relationship between behavioral assessments and ABR Wave I amplitudes appears to be more complicated. Chinchillas with selective IHC loss showed ABR thresholds, DPOAEs, and behavioral thresholds in quiet environment to be essentially unchanged despite the large IHC loss, while behavioral deficits in noise were noted [101]. Some studies have found diminished ABR Wave I amplitudes along with large TTS changes (>30 dB at 24 h post-noise), but ABR amplitude changes were not permanently diminished when TTS was smaller (<30 dB at 24 h post-noise) [54, 102, 103]. Lobarinas et al. [70] used a modified startle inhibition paradigm in rats to investigate if robust TTS and permanent reductions in suprathreshold ABR Wave I amplitudes correlated with hearing-in-noise performance. The study found that noise exposures that resulted in TTS greater than 30 dB (24 h post-noise) reduced both ABR Wave I amplitudes and hearing-in-noise performance at more difficult SNR levels. These changes were not explained by loss of audibility.

5.1   Speech-in-Noise Testing

The disconnect between the robust anatomical evidence of noise- and age-related synaptopathic losses and the lack of overt functional deficits is best summarized in recent review articles [104–107]. One of the main hypotheses of the effects of synaptopathy is poorer hearing in noise with otherwise normal hearing sensitivity. Presumably, these patients present with auditory complaints in noise without a commensurate audiometric loss. In general, studies have failed to support this hypothesis, even in the presence of abnormal temporal and intensity coding at the level of the auditory nerve. One possibility is that the speech-in-noise tasks lack sufficient sensitivity to detect the subtle effects of synaptic loss. Despite the lack of robust correlations between auditory complaints, physiological evidence, and perceptual deficits, most researchers and clinicians conclude that more research is needed and that early markers of noise-induced cochlear pathology are essential for hearing conservation efforts.

Studies have suggested that individuals with a history of noise exposure may present with suprathreshold functional deficits in spectro-temporal processing [108], difficulty understanding speech in noise [109, 110], and attentional deficits [111]. These changes could be observed before changes were noted in pure-tone thresholds, and would represent early markers for NIHL. Speech understanding assessments, especially speech-in-noise tests, are analogous to providing a "stress test" for auditory processing ability [112] and provide ecological validity to clinical and research assessment of auditory function.

Speech-in-noise deficits have been reported in noise-exposed Air Force pilots compared to nonexposed Air Force administrative personnel [109]. No differences were observed in other auditory tasks, including simultaneous masking, backward masking, and frequency discrimination paradigms. It is important to note, however, that only thresholds up to 4 kHz were reported. In contrast, findings from recreational noise exposures that divided individuals into groups based on word recognition-in-noise performance [113] showed no significant differences in preferred listening level or in noise exposure history.

Several considerations should be taken when deciding upon speech assessment and the paradigms for test administration. This includes the presentation level (i.e., 60 vs. 80 dB) and the SNR that will be sensitive enough to detect performance differences within and across listeners, while avoiding either ceiling effects (near 100% correct) for listeners with normal audibility or floor effects (chance) for listeners with hearing loss [114–116]. Currently, there is no consensus in the field about the best speech or speech-in-noise measure (i.e., WIN, QiuckSIN, the LiSN-S, SPIN, and HINT, to name a few) for noise injury detection. One factor that complicates the interpretation of the results is that the assessment variables differ across tests (e.g., variable SNR levels, talker and noise differences, and the use of words versus sentence length stimuli). In summary, noise-induced damage in humans is highly variable, and this may account for the inability to replicate animal findings in humans. It is important to establish sensitive diagnostic tools to identify early markers of hearing loss. Although some measures show potential, more studies are needed to establish functional differences following noise exposure.

6 Pharmaceutical Otoprotection from Noise-Induced Hearing Loss

Numerous otoprotective strategies have been examined to reduce NIHL, in terms of both threshold measurements and assessment of suprathreshold performance. The sections below describe some of the otoprotective compounds that have been used to prevent NIHL.

6.1 Antioxidants

Oxidative stress has been detected in the cochlea after noise exposure [16–19], and has been strongly linked to organ of Corti injuries caused by noise. Therefore, antioxidants have been identified as potential otoprotectants. R-phenylisopropyladenosine (R-PIA) increases the antioxidant glutathione and antioxidant enzyme superoxide dismutase (SOD) [117]. Glutathione scavenges the hydroxyl radical, while SOD helps convert superoxide into hydrogen peroxide, which in turn is converted into water through multiple pathways. R-PIA delivered to the round windows of chinchillas reduced PTS at 4–16 kHz and OHC loss after a 105-dBSPL, 4 kHz band noise exposure [118]. A combination of R-PIA and glutathione monoethyl ester, a glutathione precursor molecule, reduced impulse noise-induced PTS when delivered on the round windows of chinchillas [119].

Cysteine is a major component of glutathione that limits its levels in the body. Methionine has significant influence on the availability of cysteine in tissue. Therefore, increased methionine levels in the cochlea were thought to potentially increase glutathione levels, and thus exert an otoprotective effect. D-Methionine was targeted as an otoprotectant in part due to its potential availability in the cochlea after systemic administration. In mice, D-methionine was shown to reduce PTS and inhibit lipid peroxidation [120], a damaging consequence of oxidative stress. Intraperitoneal injections before and after a 6-h continuous noise exposure at 105 dB SPL significantly reduced PTS and OHC loss in chinchillas [121]. The compound also reduced PTS from the same noise exposure in chinchillas when delivered in a rescue paradigm beginning 1 h post-noise [122]. The rescue effect was also obtained when the injections started at 3, 5, or 7 h after the noise exposure, and the magnitude of protection was unchanged [123]. In the albino guinea pig, D-methionine also exerted a rescue effect when injections began 1 h after noise exposure, but the effect was dose-dependent, with the highest dose delivering the strongest rescue effect and the lowest dose delivering no effect [124].

Another compound, n-acetyl, L-cysteine (NAC) acts with the same mechanism as D-methionine by increasing glutathione through increased availability of cysteine. Local application reduced threshold shift associated with cochlear implantation [125, 126], but implantation also caused threshold shift in the high frequencies due to mechanical damage on the basal portion of the cochlea [126]. However, systemic dosing of NAC and salicylate with intraperitoneal injections before and after a 4 kHz, 105 dB SPL, 6-h continuous noise exposure provided significant protection from PTS and OHC loss compared to saline-treated controls. When the injections were given 1 h and then twice daily for 2 days after the noise exposure in a rescue paradigm, there was still protection from PTS, albeit a weaker effect [127]. Intraperitoneal NAC also provided protection against PTS and OHC loss from impulse noise [128–130] and a noise that combined continuous noise and impacts [128]. NAC was also shown to be effective as a noise otoprotectant when delivered orally in animal models [128, 131]. NAC and D-methionine were also effective in protecting against PTS in chinchillas when delivered in relatively low doses [132]. It is important to note that a similar study found that NAC provided no protection against high-level, long-duration broadband noise exposure [133].

NAC has also advanced to studies of otoprotection in humans exposed to damaging levels of noise. Taiwanese workers exposed to occupational noise levels of 88.4–89.4 dB time-weighted average were treated with NAC via oral dosing for 14 days. Workers also underwent a 2-week course of placebo treatment. The NAC treatment produced a small (<1 dB), but statistically significant, reduction in high-frequency (average of 3, 4, and 6 kHz) TTS in workers treated with NAC [134]. For NIHL from firearm use, military personnel were treated with oral NAC after shooting. Firearm exposure disrupted performance on an amplitude modulation transfer function task in control subjects, but not in the participants treated with NAC [135]. In a cross-sectional study, textile workers exposed to more than 85 dB time-weighted average continuous noise were treated with oral NAC; they developed less TTS than control subjects [136], though it should be noted that the NAC-treated participants had higher baseline thresholds than the control participants. A Phase II clinical trial ($n = 566$) in military personnel at risk for NIHL from shooting revealed no significant reductions in standard threshold shifts from noise [137]. Based on positive findings in preclinical animal studies, further experimentation in human populations seems warranted.

Ebselen is a compound that mimics the action of glutathione peroxidase [138], an enzyme that catalyzes the conversion of hydrogen peroxide into water, using glutathione as a substrate. Oral dosing with ebselen attenuated PTS and OHC loss from a 5-h noise presented at 125 dB SPL [139]. The protective effect was demonstrated in multiple animal models and with multiple noise exposures [140, 141]. In a human phase II clinical trial, participants were dosed twice daily for 4 days with oral ebselen at one of the three dose levels: 200, 400, or 600 mg/kg. They were then exposed to a TTS-inducing noise, and TTS was measured 15 min later. TTS was significantly (>50%) reduced in the group treated at the 400 mg/kg dose, but not the 200 or 600 mg/kg dosing levels [142].

Salicylate and Trolox, an analog of vitamin E, exerted rescue effects against NIHL when given in combination, but not separately. Salicylate has free radical scavenging properties, and vitamin E scavenges the peroxyl radical [143]. The series of injections began at different times after noise exposure, and this combined treatment significantly reduced PTS, OHC loss, and oxidative stress markers. The rescue effect was strongest when the time interval between the noise and the first injection was shortest, and weakest with the longest interval [144]. Vitamin E was also effective in reducing PTS and OHC loss when given in combination with vitamins A and C and magnesium. It is notable that the combination of vitamins A, C, and E did not provide protection without magnesium [25]. This combination was also effective when given chronically as a dietary supplement. Mice fed with a diet enhanced with vitamins A, C, E, and magnesium had lower PTS and less lateral wall injury after a noise exposure [145].

6.2 Steroid Therapy

Steroid treatment is the standard of care for disease management and treatment of sudden sensorineural hearing loss, and steroids have undergone considerable evaluation as otoprotective agents for NIHL. Direct administration of dexamethasone, an anti-inflammatory corticosteroid, into the inner ear and intravenously administered dehydroepiandrosterone, a neurosteroid hormone, has shown potential therapeutic effects on NIHL in animal models when given before [146–148] or after [149] acoustic trauma. Beneficial treatment effects were seen for several different doses; however higher doses appear to be associated with better hearing preservation [146, 147]. The therapeutic time window for treatment appears to be very short [150]. Nevertheless, evidence of positive treatment effects is still unclear as another study did not find NIHL protective effects of dexamethasone [151].

Several steroids (i.e., pregnenolone and progesterone) can be synthesized within the central nervous system and are referred to as "neurosteroids." Dehydroepiandrosterone sulfate (DHEAS) is a neurosteroid that is secreted by the adrenal cortex in response to adrenocorticotrophine. DHEAS has antioxidant and neuroprotective properties and has protective effects against some brain injuries, including those from glutamate excitotoxicity and anoxia–reoxygenation [152–154]. Tabuchi et al. [147] analyzed the effects of DHEAS on acoustic injury in an animal model. Albino guinea pigs were administered DHEAS (each animal in the experimental group were given 0.1 or 1 mg/kg) or a saline solution intravenously and were then immediately exposed to a 2-kHz pure tone of 120 or 125 dB SPL for 10 min. Immediately after the acoustic overexposure, DHEAS did not show protective effects (as measured by the compound action potential [CAP] threshold or DPOAE amplitudes). However, statistically significant improvements in compound CAP thresholds and DPOAE amplitudes were observed 1 week following exposure in the animals treated with the higher dose of DHEAS for both noise levels (120 and 125 dB SPL). Protective effects were not observed for the lower, 0.1 mg/kg dosage. These results suggest that DHEAS has a protective effect against NIHL [147].

Dexamethasone, a corticosteroid commonly used for its anti-inflammatory properties, was also evaluated for potential protective effects against noise-induced trauma in guinea pigs [148]. Animals were administered various doses of dexamethasone or the control (artificial perilymph) via a mini-osmotic pump placed directly into the scala tympani. Four days following drug delivery, animals were exposed to a 120-dB SPL octave band noise, centered at 4 kHz for 24 h. The study reported a dose-dependent reduction in hair cell death and better ABR thresholds in the animals that received dexamethasone compared to the control animals following noise overexposure. Findings suggest protective cochlear effects of dexamethasone when infused directly into the perilymphatic space [148].

A comparative study [151] evaluated the effectiveness of dexamethasone as well as melatonin, an antioxidant, and tacrolimus, an immunosuppressant agent, against NIHL in 30 Wistar rats. Rats were exposed to white noise at 120 dB SPL for 4 h. Each animal received one of the three investigational compounds the day before noise exposure and during the following 14 days postexposure. Functional protective effects were determined by DPOAE amplitudes, ABR thresholds, cytocochleograms, and gene expression analysis at day 21 postexposure. Bas et al. [151] concluded that dexamethasone did not have otoprotective effects from noise overexposure, while animals treated with melatonin and tacrolimus had better functional hearing outcomes suggesting that these compounds were otoprotective against NIHL. The study did not have a control comparison group, which greatly limits the interpretation of these results. Thus, further studies are needed to understand the molecular mechanisms involved and to confirm the otoprotective properties of these compounds.

Psillas et al. [155] evaluated the effect of early combined treatment of steroids and piracetam (a derivative of GABA that has neuroprotective and anticonvulsant properties) for acute acoustic trauma in 52 young-adult soldiers. A proposed mechanism of acute acoustic trauma is cochlear hypoxia, which can occur after intense acoustic exposures, such as gunfire. Hearing improvement following the combined treatment was observed in 69% of participants. However, it is difficult to determine the prevalence of spontaneous recovery as there was no control group for comparisons. Patients that received earlier treatment had significantly better hearing 1 week following acoustic trauma; this was in agreement with previous research that indicated hearing outcomes were greatly dependent on the number of days before the start of treatment following acute acoustic trauma [156]. Further randomized control trials with larger sample sizes are needed to draw conclusions on the combined treatment with steroids and piracetam. However, this study indicates the potential importance of immediate intervention needed for acute acoustic trauma.

A systematic review was completed to evaluate treatment options and hearing outcomes between acoustic trauma and other types of acute NIHL [157]. Most of the cases reviewed by the researchers included military personnel that were exposed to firearms and explosions, and were regarded as acoustic trauma insults. Systemic steroid therapy was the treatment option most frequently used among the studies and cases reviewed. However, the efficacy of systemic steroid use was unclear across the studies reviewed and recovery was only observed in those within the acute NIHL group. Wada et al. [157] suggested that cochlear damage due to acoustic trauma and the efficacy of systemic steroid therapy might depend on the severity of the cochlear damage following acoustic trauma, presumably because severe mechanical damage to the cochlea may be untreatable. Thus, it may be important to differentiate between the type and severity of acoustic trauma to gauge the efficacy of hearing protection in clinical settings and in the trials evaluating treatment effects.

Studies suggest that steroid treatment may be effective for the management of acute acoustic injuries [157], but high-quality randomized controlled trials on the efficacy of steroids as an otoprotective agent are lacking. It is important to note that a completely randomized study on steroid treatment is difficult to perform due to ethical considerations. Experimental noise conditions varied considerably among the studies reviewed here, limiting the interpretation of the otoprotective results in cases of NIHL. Finally, most studies on steroids lack complete dose-response curves from which it would be possible to identify the most otoprotective dose to be administered before, during, or after a noise exposure. Therefore, further studies are needed to clarify the efficacy of steroids as otoprotective agents against various type of noise exposure.

6.3   S-Ketamine and Glutamate Excitotoxicity

Ketamine exists as two isomer forms: the racemic, or $R(-)$, and the S-enantiomer, or $S(+)$, ketamine. S-Ketamine is thought to have neuroprotective and regenerative effects [158–160]. Recently, ketamine has attracted interest as a potential therapeutic intervention for brain injury in preclinical studies. Findings indicate a reduction of neural damage in the cortex by mitigating the excitotoxic effects of glutamate in vitro [161] and reducing neuronal damage in rats experiencing cerebral ischemia [159, 162, 163] or head trauma [164]. This protective mechanism is believed to involve

N-methyl-D-aspartic acid (NMDA) and non-NMDA receptors. Local administration of AM-101, an NMDA receptor antagonist, into the cochlea reduced the level on noise-induced trauma to the IHCs [165]. S-Ketamine has the benefit of having a rapid onset and short duration of action. Furthermore, it does not seem to affect cerebral autoregulation of oxygen and glucose metabolic rate when administered during Propofol anesthesia [166, 167]. In the cochlea, NMDA receptors have been found at the IHC postsynaptic terminals and are upregulated during glutamate excitotoxicity [168, 169]. This excitotoxic effect can be triggered by acoustic trauma or hypoxia [56], suggesting that noise-induced cochlear damage may involve NMDA receptor activity. Because ketamine is an NMDA receptor antagonist, it has been suggested as a treatment to block noise-induced glutamate excitotoxicity. Giraudet et al. [38] have found protective effects in rats anesthetized with ketamine and xylazine with lower TTS levels relative to awake animals. The study suggests that reduced glutamate excitotoxicity provided by the ketamine may be otoprotective.

6.4   Cell Death and Stress Inhibitors

The apoptosis pathway (see Sect. 2.3) is an extremely complex, multifaceted sequence of signaling events that culminates in cell death. Initial signaling can occur from the cell membrane, the mitochondria, or the nucleus, and the source of stress can come from multiple locations inside and outside of the cell. Thus, there are numerous points in the cascade of apoptotic events where intervention could potentially lead to a cell survival outcome instead of a cell death outcome.

The first major series of experiments into otoprotection through inhibition of apoptosis signaling was with c-Jun N-terminal kinase (JNK), which has been implicated in cochlear injury from cochlear implant insertion trauma [170, 171], ototoxic drugs [172, 173] and noise [172, 174, 175]. JNK is a key molecule in multiple apoptosis pathways through its role in stress-activated protein kinase pathways [176]. JNK inhibitors have been shown to exert otoprotective effects against continuous noise exposures. Local application of D-JNK-1, a cell-permeable JNK, exerted an otoprotective effect when delivered via an implanted osmotic pump directly into scala tympani 30 min prior to noise exposure [172]. Local dosing of D-JNK-1 also worked in a rescue paradigm, with decreasing efficacy as the time between the noise exposure and onset of rescue increased. Intervention needed to begin no later than 6 h after the exposure in order to provide some rescue effect [175]. The D-JNK-1 inhibitor also improved recovery from impulse noise exposures when delivered in a rescue paradigm. Local delivery into the cochlea through an implanted osmotic pump or hyaluronic acid gel on the round window was more effective than systemic intraperitoneal though all delivery approaches did improve recovery from the noise compared to controls [177]. CEP-1347, another JNK inhibitor, attenuated PTS and hair cell loss from a 6-h, 120 dB SPL noise when given for a 2-week period starting 4 h before the noise via subcutaneous injections [174], thus supporting the efficacy of JNK inhibitors seen with systemic drug delivery.

Src protein-tyrosine kinase (PTK) is involved in a very broad set of cellular actions, including cell migration, cell proliferation, and cell death through apoptosis. Src PTKs are involved in signaling between focal adhesions on the cell's membrane and the cell nucleus; this provides the nucleus with information about the cell's physical stability [178]. Src was targeted for otoprotection from noise because it has a role in ROS formation, but it was also hypothesized to play a role in anoikis [179], which is rapid apoptosis induced by mechanical trauma to the cell or its extracellular matrix [180]. Inhibition of Src with a series of compounds delivered locally onto the round window of chinchillas reduced PTS and hair cell loss from continuous and impulse noises [179, 181]. The most effective protective compound, KX1-004, also reduced PTS in chinchillas when administered subcutaneously both before a 6-h continuous noise repeated for 4 consecutive days [130, 182] or when administered in a rescue approach after a 1-h continuous noise exposure at 112-dB SPL [183].

Pifithrin-α is an inhibitor of p53, which is a tumor suppressor involved in apoptosis (see Miller et al. [184] for a review of the many roles of p53 in developmental and acquired cell death). Genes related to p53 are expressed at higher levels after noise exposure [185]. When delivered to the round window prior to exposure, pifithrin-α provided protection from impulse noise-induced compound threshold shift measured at 3 days after exposure [186].

7 Summary

Although the experiments described provide optimism about the ability to pharmaceutically intercede in various cell death pathways, there is considerable uncertainty about the ability to translate the positive results observed in animal models into human clinical applications. Further, because the apoptosis signaling cascades are so complex, it is unclear what the implications are of intervening in one pathway, but not others. The positive otoprotection results described above represent considerable proof of concept, but further refinement is still needed to optimize the effectiveness and safety of these compounds and evaluate their efficacy in noise-exposed humans.

References

1. Nelson DI, Nelson RY, Concha-Barrientos M, Fingerhut M (2005) The global burden of occupational noise-induced hearing loss. Am J Ind Med 48(6):446–458. https://doi.org/10.1002/ajim.20223
2. Henderson E, Testa MA, Hartnick C (2011) Prevalence of noise-induced hearing-threshold shifts and hearing loss among US youths. Pediatrics 127(1):e39–e46. https://doi.org/10.1542/peds.2010-0926

3. Chiu LL, Cunningham LL, Raible DW, Rubel EW, Ou HC (2008) Using the zebrafish lateral line to screen for ototoxicity. J Assoc Res Otolaryngol 9(2):178–190. https://doi.org/10.1007/s10162-008-0118-y

4. Ou HC, Raible DW, Rubel EW (2007) Cisplatin-induced hair cell loss in zebrafish (Danio rerio) lateral line. Hear Res 233(1–2):46–53. https://doi.org/10.1016/j.heares.2007.07.003

5. Owens KN, Coffin AB, Hong LS, Bennett KO, Rubel EW, Raible DW (2009) Response of mechanosensory hair cells of the zebrafish lateral line to aminoglycosides reveals distinct cell death pathways. Hear Res 253(1–2):32–41. https://doi.org/10.1016/j.heares.2009.03.001

6. Ding D, He J, Allman BL, Yu D, Jiang H, Seigel GM, Salvi RJ (2011) Cisplatin ototoxicity in rat cochlear organotypic cultures. Hear Res 282(1–2):196–203. https://doi.org/10.1016/j.heares.2011.08.002

7. Sha SH, Taylor R, Forge A, Schacht J (2001) Differential vulnerability of basal and apical hair cells is based on intrinsic susceptibility to free radicals. Hear Res 155(1–2):1–8

8. Feghali JG, Liu W, Van De Water TR (2001) L-n-acetyl-cysteine protection against cisplatin-induced auditory neuronal and hair cell toxicity. Laryngoscope 111(7):1147–1155. https://doi.org/10.1097/00005537-200107000-00005

9. Staecker H, Liu W, Malgrange B, Lefebvre PP, Van De Water TR (2007) Vector-mediated delivery of bcl-2 prevents degeneration of auditory hair cells and neurons after injury. ORL J Otorhinolaryngol Relat Spec 69(1):43–50. https://doi.org/10.1159/000096716

10. Henderson D, Bielefeld EC, Harris KC, Hu BH (2006) The role of oxidative stress in noise-induced hearing loss. Ear Hear 27(1):1–19. https://doi.org/10.1097/01.aud.0000191942.36672.f3

11. Ohlemiller KK, Rice ME, Gagnon PM (2008) Strial microvascular pathology and age-associated endocochlear potential decline in NOD congenic mice. Hear Res 244(1–2):85–97. https://doi.org/10.1016/j.heares.2008.08.001

12. Wang Y, Hirose K, Liberman MC (2002) Dynamics of noise-induced cellular injury and repair in the mouse cochlea. J Assoc Res Otolaryngol 3(3):248–268. https://doi.org/10.1007/s101620020028

13. Harding GW, Bohne BA, Ahmad M (2002) DPOAE level shifts and ABR threshold shifts compared to detailed analysis of histopathological damage from noise. Hear Res 174(1–2):158–171

14. Nordmann AS, Bohne BA, Harding GW (2000) Histopathological differences between temporary and permanent threshold shift. Hear Res 139(1–2):13–30

15. Santi PA, Duvall AJ 3rd. (1978) Stria vascularis pathology and recovery following noise exposure. Otolaryngology 86(2):ORL354–ORL361

16. Ohinata Y, Miller JM, Altschuler RA, Schacht J (2000) Intense noise induces formation of vasoactive lipid peroxidation products in the cochlea. Brain Res 878(1–2):163–173

17. Ohlemiller KK, Wright JS, Dugan LL (1999) Early elevation of cochlear reactive oxygen species following noise exposure. Audiol Neurootol 4(5):229–236. https://doi.org/10.1159/000013846

18. Yamane H, Nakai Y, Takayama M, Iguchi H, Nakagawa T, Kojima A (1995) Appearance of free radicals in the guinea pig inner ear after noise-induced acoustic trauma. Eur Arch Otorhinolaryngol 252(8):504–508

19. Yamashita D, Jiang HY, Schacht J, Miller JM (2004) Delayed production of free radicals following noise exposure. Brain Res 1019(1–2):201–209. https://doi.org/10.1016/j.brainres.2004.05.104

20. Puel JL, Ruel J, Gervais d'Aldin C, Pujol R (1998) Excitotoxicity and repair of cochlear synapses after noise-trauma induced hearing loss. Neuroreport 9(9):2109–2114

21. Hu BH, Guo W, Wang PY, Henderson D, Jiang SC (2000) Intense noise-induced apoptosis in hair cells of guinea pig cochleae. Acta Otolaryngol 120(1):19–24

22. Lefebvre PP, Malgrange B, Lallemend F, Staecker H, Moonen G, Van De Water TR (2002) Mechanisms of cell death in the injured auditory system: otoprotective strategies. Audiol Neurootol 7(3):165–170. https://doi.org/10.1159/000058304

23. Nakagawa T, Yamane H, Shibata S, Takayama M, Sunami K, Nakai Y (1997) Two modes of auditory hair cell loss following acoustic overstimulation in the avian inner ear. ORL J Otorhinolaryngol Relat Spec 59(6):303–310. https://doi.org/10.1159/000276961

24. Nicotera TM, Hu BH, Henderson D (2003) The caspase pathway in noise-induced apoptosis of the chinchilla cochlea. J Assoc Res Otolaryngol 4(4):466–477. https://doi.org/10.1007/s10162-002-3038-2

25. Le Prell CG, Hughes LF, Miller JM (2007) Free radical scavengers vitamins A, C, and E plus magnesium reduce noise trauma. Free Radic Biol Med 42(9):1454–1463. https://doi.org/10.1016/j.freeradbiomed.2007.02.008

26. Le Prell CG, Yamashita D, Minami SB, Yamasoba T, Miller JM (2007) Mechanisms of noise-induced hearing loss indicate multiple methods of prevention. Hear Res 226(1–2):22–43. https://doi.org/10.1016/j.heares.2006.10.006

27. Chen YS, Tseng FY, Liu TC, Lin-Shiau SY, Hsu CJ (2005) Involvement of nitric oxide generation in noise-induced temporary threshold shift in guinea pigs. Hear Res 203(1–2):94–100. https://doi.org/10.1016/j.heares.2004.12.006

28. Diao M, Gao W, Sun J (2007) Nitric oxide synthase inhibitor reduces noise-induced cochlear damage in guinea pigs. Acta Otolaryngol 127(11):1162–1167. https://doi.org/10.1080/00016480701242436

29. Bielefeld EC (2013) Reduction in impulse noise-induced permanent threshold shift with intracochlear application of an NADPH oxidase inhibitor. J Am Acad Audiol 24(6):461–473. https://doi.org/10.3766/jaaa.24.6.3

30. Bielefeld EC, Hu BH, Harris KC, Henderson D (2005) Damage and threshold shift resulting from cochlear exposure to paraquat-generated superoxide. Hear Res 207(1–2):35–42. https://doi.org/10.1016/j.heares.2005.03.025

31. Ramkumar V, Whitworth CA, Pingle SC, Hughes LF, Rybak LP (2004) Noise induces A1 adenosine receptor expression in the chinchilla cochlea. Hear Res 188(1–2):47–56. https://doi.org/10.1016/S0378-5955(03)00344-7

32. Bottger EC, Schacht J (2013) The mitochondrion: a perpetrator of acquired hearing loss. Hear Res 303:12–19. https://doi.org/10.1016/j.heares.2013.01.006

33. Miller JM, Dengerink H (1988) Control of inner ear blood flow. Am J Otolaryngol 9(6):302–316

34. Laurikainen EA, Kim D, Didier A, Ren T, Miller JM, Quirk WS, Nuttall AL (1993) Stellate ganglion drives sympathetic regulation of cochlear blood flow. Hear Res 64(2):199–204

35. Ren TY, Laurikainen E, Quirk WS, Miller JM, Nuttall AL (1993) Effects of electrical stimulation of the superior cervical ganglion on cochlear blood flow in guinea pig. Acta Otolaryngol 113(2):146–151

36. Bielefeld EC, Henderson D (2007) Influence of sympathetic fibers on noise-induced hearing loss in the chinchilla. Hear Res 223(1–2):11–19. https://doi.org/10.1016/j.heares.2006.09.010

37. Borg E (1982) Susceptibility of the sympathectomized ear to noise-induced hearing loss. Acta Physiol Scand 114(3):387–391. https://doi.org/10.1111/j.1748-1716.1982.tb06999.x

38. Giraudet F, Horner KC, Cazals Y (2002) Similar half-octave TTS protection of the cochlea by xylazine/ketamine or sympathectomy. Hear Res 174(1–2):239–248

39. Hildesheimer M, Henkin Y, Pye A, Heled S, Sahartov E, Shabtai EL, Muchnik C (2002) Bilateral superior cervical sympathectomy and noise-induced, permanent threshold shift in guinea pigs. Hear Res 163(1–2):46–52

40. Hildesheimer M, Sharon R, Muchnik C, Sahartov E, Rubinstein M (1991) The effect of bilateral sympathectomy on noise induced temporary threshold shift. Hear Res 51(1):49–53

41. Lamm K, Arnold W (2000) The effect of blood flow promoting drugs on cochlear blood flow, perilymphatic pO(2) and auditory function in the normal and noise-damaged hypoxic and ischemic guinea pig inner ear. Hear Res 141(1–2):199–219

42. Miller JM, Brown JN, Schacht J (2003) 8-Iso-prostaglandin F(2alpha), a product of noise exposure, reduces inner ear blood flow. Audiol Neurootol 8(4):207–221. https://doi.org/10.1159/000071061

43. Perlman HB, Kimura R (1962) Cochlear blood flow in acoustic trauma. Acta Otolaryngol 54:99–110
44. Kerr JF, Wyllie AH, Currie AR (1972) Apoptosis: a basic biological phenomenon with wide-ranging implications in tissue kinetics. Br J Cancer 26(4):239–257
45. Sokolova IM, Evans S, Hughes FM (2004) Cadmium-induced apoptosis in oyster hemocytes involves disturbance of cellular energy balance but no mitochondrial permeability transition. J Exp Biol 207(Pt 19):3369–3380. https://doi.org/10.1242/jeb.01152
46. Hu BH, Zheng GL (2008) Membrane disruption: an early event of hair cell apoptosis induced by exposure to intense noise. Brain Res 1239:107–118. https://doi.org/10.1016/j.brainres.2008.08.043
47. Van De Water TR, Lallemend F, Eshraghi AA, Ahsan S, He J, Guzman J et al (2004) Caspases, the enemy within, and their role in oxidative stress-induced apoptosis of inner ear sensory cells. Otol Neurotol 25(4):627–632
48. Hu BH, Henderson D, Nicotera TM (2002) Involvement of apoptosis in progression of cochlear lesion following exposure to intense noise. Hear Res 166(1–2):62–71
49. Hu BH, Henderson D, Nicotera TM (2006) Extremely rapid induction of outer hair cell apoptosis in the chinchilla cochlea following exposure to impulse noise. Hear Res 211(1–2):16–25. https://doi.org/10.1016/j.heares.2005.08.006
50. Slepecky N, Hamernik R, Henderson D, Coling D (1981) Ultrastructural changes to the cochlea resulting from impulse noise. Arch Otorhinolaryngol 230(3):273–278
51. Ahmad M, Bohne BA, Harding GW (2003) An in vivo tracer study of noise-induced damage to the reticular lamina. Hear Res 175(1–2):82–100
52. Salvi RJ, Hamernik RP, Henderson D (1979) Auditory nerve activity and cochlear morphology after noise exposure. Arch Otorhinolaryngol 224(1–2):111–116
53. Henderson D, Salvi R, Pavek G, Hamernik R (1984) Amplitude modulation thresholds in chinchillas with high-frequency hearing loss. J Acoust Soc Am 75(4):1177–1183
54. Fernandez KA, Jeffers PW, Lall K, Liberman MC, Kujawa SG (2015) Aging after noise exposure: acceleration of cochlear synaptopathy in "recovered"ears. J Neurosci 35(19):7509–7520. https://doi.org/10.1523/JNEUROSCI.5138-14.2015
55. Kujawa SG, Liberman MC (2009) Adding insult to injury: cochlear nerve degeneration after "temporary" noise-induced hearing loss. J Neurosci 29(45):14077–14085. https://doi.org/10.1523/JNEUROSCI.2845-09.2009
56. Pujol R, Puel JL (1999) Excitotoxicity, synaptic repair, and functional recovery in the mammalian cochlea: a review of recent findings. Ann N Y Acad Sci 884:249–254
57. Robertson D (1983) Functional significance of dendritic swelling after loud sounds in the guinea pig cochlea. Hear Res 9(3):263–278
58. Furman AC, Kujawa SG, Liberman MC (2013) Noise-induced cochlear neuropathy is selective for fibers with low spontaneous rates. J Neurophysiol 110(3):577–586. https://doi.org/10.1152/jn.00164.2013
59. Makary CA, Shin J, Kujawa SG, Liberman MC, Merchant SN (2011) Age-related primary cochlear neuronal degeneration in human temporal bones. J Assoc Res Otolaryngol 12(6):711–717. https://doi.org/10.1007/s10162-011-0283-2
60. Viana LM, O'Malley JT, Burgess BJ, Jones DD, Oliveira CA, Santos F et al (2015) Cochlear neuropathy in human presbycusis: confocal analysis of hidden hearing loss in post-mortem tissue. Hear Res 327:78–88. https://doi.org/10.1016/j.heares.2015.04.014
61. Kujawa SG, Liberman MC (2015) Synaptopathy in the noise-exposed and aging cochlea: primary neural degeneration in acquired sensorineural hearing loss. Hear Res 330(Pt B):191–199. https://doi.org/10.1016/j.heares.2015.02.009
62. Parthasarathy A, Kujawa SG (2018) Synaptopathy in the aging cochlea: characterizing early-neural deficits in auditory temporal envelope processing. J Neurosci 38(32):7108–7119. https://doi.org/10.1523/JNEUROSCI.3240-17.2018

63. Sergeyenko Y, Lall K, Liberman MC, Kujawa SG (2013) Age-related cochlear synaptopathy: an early-onset contributor to auditory functional decline. J Neurosci 33(34):13686–13694. https://doi.org/10.1523/JNEUROSCI.1783-13.2013

64. Bourien J, Tang Y, Batrel C, Huet A, Lenoir M, Ladrech S, Desmadryl G, Nouvian R, Puel JL, Wang J (2014) Contribution of auditory nerve fibers to compound action potential of the auditory nerve. J Neurophysiol. 112(5):1025–1039. https://doi.org/10.1152/jn.00738.2013

65. Lin HW, Furman AC, Kujawa SG, Liberman MC (2011) Primary neural degeneration in the Guinea pig cochlea after reversible noise-induced threshold shift. J Assoc Res Otolaryngol 12(5):605–616. https://doi.org/10.1007/s10162-011-0277-0

66. Gleich O, Semmler P, Strutz J (2016) Behavioral auditory thresholds and loss of ribbon synapses at inner hair cells in aged gerbils. Exp Gerontol 84:61–70. https://doi.org/10.1016/j.exger.2016.08.011

67. Mohrle D, Ni K, Varakina K, Bing D, Lee SC, Zimmermann U et al (2016) Loss of auditory sensitivity from inner hair cell synaptopathy can be centrally compensated in the young but not old brain. Neurobiol Aging 44:173–184. https://doi.org/10.1016/j.neurobiolaging.2016.05.001

68. Hickman TT, Smalt C, Bobrow J, Quatieri T, Liberman MC (2018) Blast-induced cochlear synaptopathy in chinchillas. Sci Rep 8(1):10740. https://doi.org/10.1038/s41598-018-28924-7

69. Valero MD, Burton JA, Hauser SN, Hackett TA, Ramachandran R, Liberman MC (2017) Noise-induced cochlear synaptopathy in rhesus monkeys (Macaca mulatta). Hear Res 353:213–223. https://doi.org/10.1016/j.heares.2017.07.003

70. Lobarinas E, Spankovich C, Le Prell CG (2017) Evidence of "hidden hearing loss" following noise exposures that produce robust TTS and ABR wave-I amplitude reductions. Hear Res 349:155–163. https://doi.org/10.1016/j.heares.2016.12.009

71. Forster J, Wendler O, Buchheidt-Doerfler I, Krauss P, Schilling A, Sterna E, Schulze H, Tziridis K (2018) Tinnitus development is associated with synaptopathy of inner hair cells in Mongolian gerbils. Preprint. https://doi.org/10.1101/304576

72. Pienkowski M (2018) Prolonged exposure of CBA/Ca mice to moderately loud noise can cause cochlear synaptopathy but not tinnitus or hyperacusis as assessed with the acoustic startle reflex. Trends Hear 22:2331216518758109. https://doi.org/10.1177/2331216518758109

73. Peters EN (1965) Temporary shifts in auditory thresholds of chinchilla after exposure to noise. J Acoust Soc Am 37:831–833

74. Hamernik RP, Ahroon WA, Patterson JH Jr, Qiu W (2002) Relations among early postexposure noise-induced threshold shifts and permanent threshold shifts in the chinchilla. J Acoust Soc Am 111(1 Pt 1):320–326

75. Henderson D, Hamernik RP, Sitler RW (1974) Audiometric and histological correlates of exposure to 1-msec noise impulses in the chinchilla. J Acoust Soc Am 56(4):1210–1221

76. Mills JH (1973) Temporary and permanent threshold shifts produced by nine-day exposures to noise. J Speech Hear Res 16(3):426–438

77. Saunders JC, Mills JH, Miller JD (1977) Threshold shift in the chinchilla from daily exposure to noise for six hours. J Acoust Soc Am 61(2):558–570

78. Ryan A, Bone RC (1978) Noise-induced threshold shift and cochlear pathology in the Mongolian gerbil. J Acoust Soc Am 63(4):1145–1151

79. Ryan AF, Kujawa SG, Hammill T, Le Prell C, Kil J (2016) Temporary and permanent noise-induced threshold shifts: a review of basic and clinical observations. Otol Neurotol 37(8):e271–e275. https://doi.org/10.1097/MAO.0000000000001071

80. Rosowski JJ (1991) The effects of external- and middle-ear filtering on auditory threshold and noise-induced hearing loss. J Acoust Soc Am 90(1):124–135

81. Clark WW, Bohne BA (1978) Animal model for the 4-kHz tonal dip. Ann Otol Rhinol Laryngol Suppl 87(4 Pt 2 Suppl 51):1–16

82. Lie A, Engdahl B, Hoffman HJ, Li CM, Tambs K (2017) Occupational noise exposure, hearing loss, and notched audiograms in the HUNT Nord-Trondelag hearing loss study, 1996-1998. Laryngoscope 127(6):1442–1450. https://doi.org/10.1002/lary.26256

83. Hamernik RP, Patterson JH, Turrentine GA, Ahroon WA (1989) The quantitative relation between sensory cell loss and hearing thresholds. Hear Res 38(3):199–211
84. Somma G, Pietroiusti A, Magrini A, Coppeta L, Ancona C, Gardi S et al (2008) Extended high-frequency audiometry and noise induced hearing loss in cement workers. Am J Ind Med 51(6):452–462. https://doi.org/10.1002/ajim.20580
85. Riga M, Korres G, Balatsouras D, Korres S (2010) Screening protocols for the prevention of occupational noise-induced hearing loss: the role of conventional and extended high frequency audiometry may vary according to the years of employment. Med Sci Monit 16(7):CR352–CR356
86. Ahmed HO, Dennis JH, Badran O, Ismail M, Ballal SG, Ashoor A, Jerwood D (2001) High-frequency (10-18 kHz) hearing thresholds: reliability, and effects of age and occupational noise exposure. Occup Med (Lond) 51(4):245–258
87. Borchgrevink HM, Hallmo P, Mair IWS (1996) Extended highfrequency hearing loss from noise exposure. In: Axelsson A, Borchgrevink HM, Hamernik RP, Hellstrom PA, Henderson D, Salvi RJ (eds), Scientific basis of noise-induced hearing loss. Thieme, Leipzig, Germany. pp 299–312
88. Le Prell CG, Bao J (2012) Prevention of noise-induced hearing loss: potential therapeutic agents. In: Le Prell HD, CG FRR, Popper AN (eds) Noise-induced hearing loss: scientific advances, Springer handbook of auditory research. Springer Science+Business Media, LLC, New York, pp 285–338
89. Le Prell CG, Lobarinas E (2015) Strategies for assessing antioxidant efficacy in clinical trials. In: Miller JM, Le Prell CG, Rybak L (eds) Oxidative stress in applied basic research and clinical practice: free radicals in ENT pathology. Humana Press, New York, pp 163–192
90. Stamper GC, Johnson TA (2015) Auditory function in normal-hearing, noise-exposed human ears. Ear Hear 36(2):172–184. https://doi.org/10.1097/AUD.0000000000000107
91. Stamper GC, Johnson TA (2015) Letter to the Editor: Examination of potential sex influences in auditory function in normal-hearing, noise-exposed human ears, Ear Hear 36, 172–184. Ear Hear 36(6):738–740. https://doi.org/10.1097/AUD.0000000000000228
92. Fulbright ANC, Le Prell CG, Griffiths SK, Lobarinas E (2017) Effects of recreational noise on threshold and suprathreshold measures of auditory function. Semin Hear 38(4):298–318. https://doi.org/10.1055/s-0037-1606325
93. Grinn SK, Wiseman KB, Baker JA, Le Prell CG (2017) Hidden hearing loss? No effect of common recreational noise exposure on cochlear nerve response amplitude in humans. Front Neurosci 11:465. https://doi.org/10.3389/fnins.2017.00465
94. Liberman MC, Epstein MJ, Cleveland SS, Wang H, Maison SF (2016) Toward a differential diagnosis of hidden hearing loss in humans. PLoS One 11(9):e0162726. https://doi.org/10.1371/journal.pone.0162726
95. Valderrama JT, Beach EF, Yeend I, Sharma M, Van Dun B, Dillon H (2018) Effects of lifetime noise exposure on the middle-age human auditory brainstem response, tinnitus and speech-in-noise intelligibility. Hear Res 365:36–48. https://doi.org/10.1016/j.heares.2018.06.003
96. Bramhall NF, Konrad-Martin D, McMillan GP, Griest SE (2017) Auditory brainstem response altered in humans with noise exposure despite normal outer hair cell function. Ear Hear 38(1):e1–e12. https://doi.org/10.1097/AUD.0000000000000370
97. Prendergast G, Guest H, Munro KJ, Kluk K, Leger A, Hall DA et al (2017) Effects of noise exposure on young adults with normal audiograms I: electrophysiology. Hear Res 344:68–81. https://doi.org/10.1016/j.heares.2016.10.028
98. Guest H, Munro KJ, Prendergast G, Howe S, Plack CJ (2017) Tinnitus with a normal audiogram: relation to noise exposure but no evidence for cochlear synaptopathy. Hear Res 344:265–274. https://doi.org/10.1016/j.heares.2016.12.002
99. Musiek FE, Kibbe K, Rackliffe L, Weider DJ (1984) The auditory brain stem response I-V amplitude ratio in normal, cochlear, and retrocochlear ears. Ear Hear 5(1):52–55
100. Mehraei G, Hickox AE, Bharadwaj HM, Goldberg H, Verhulst S, Liberman MC, Shinn-Cunningham BG (2016) Auditory brainstem response latency in noise as a marker

of cochlear synaptopathy. J Neurosci 36(13):3755–3764. https://doi.org/10.1523/JNEUROSCI.4460-15.2016

101. Lobarinas E, Salvi R, Ding D (2016) Selective inner hair cell dysfunction in chinchillas impairs hearing-in-noise in the absence of outer hair cell loss. J Assoc Res Otolaryngol 17(2):89–101. https://doi.org/10.1007/s10162-015-0550-8

102. Hickox AE, Liberman MC (2014) Is noise-induced cochlear neuropathy key to the generation of hyperacusis or tinnitus? J Neurophysiol 111(3):552–564. https://doi.org/10.1152/jn.00184.2013

103. Jensen JB, Lysaght AC, Liberman MC, Qvortrup K, Stankovic KM (2015) Immediate and delayed cochlear neuropathy after noise exposure in pubescent mice. PLoS One 10(5):e0125160. https://doi.org/10.1371/journal.pone.0125160

104. Barbee CM, James JA, Park JH, Smith EM, Johnson CE, Clifton S, Danhauer JL (2018) Effectiveness of auditory measures for detecting hidden hearing loss and/or cochlear synaptopathy: a systematic review. Semin Hear 39(2):172–209. https://doi.org/10.1055/s-0038-1641743

105. Chen GD (2018) Hidden cochlear impairments. J Otol 13(2):37–43. https://doi.org/10.1016/j.joto.2018.05.001

106. Kobel M, Le Prell CG, Liu J, Hawks JW, Bao J (2017) Noise-induced cochlear synaptopathy: past findings and future studies. Hear Res 349:148–154. https://doi.org/10.1016/j.heares.2016.12.008

107. Shi L, Chang Y, Li X, Aiken S, Liu L, Wang J (2016) Cochlear synaptopathy and noise-induced hidden hearing loss. Neural Plast 2016:6143164. https://doi.org/10.1155/2016/6143164

108. Stone MA, Moore BC (2008) Effects of spectro-temporal modulation changes produced by multi-channel compression on intelligibility in a competing-speech task. J Acoust Soc Am 123(2):1063–1076. https://doi.org/10.1121/1.2821969

109. Hope AJ, Luxon LM, Bamiou DE (2013) Effects of chronic noise exposure on speech-in-noise perception in the presence of normal audiometry. J Laryngol Otol 127(3):233–238. https://doi.org/10.1017/S002221511200299X

110. Suting BM (2016) Assessing the knowledge regarding environmental sanitation and its impact on health among the people in a selected rural community of Meghalaya. Nurs J India 107(4):153–155

111. Kujala T, Shtyrov Y, Winkler I, Saher M, Tervaniemi M, Sallinen M et al (2004) Long-term exposure to noise impairs cortical sound processing and attention control. Psychophysiology 41(6):875–881. https://doi.org/10.1111/j.1469-8986.2004.00244.x

112. Wilson RH (2011) Clinical experience with the words-in-noise test on 3430 veterans: comparisons with pure-tone thresholds and word recognition in quiet. J Am Acad Audiol 22(7):405–423. https://doi.org/10.3766/jaaa.22.7.3

113. Le Prell CG, Lobarinas E (2016) Clinical and translational research: challenges to the field. In: Le Prell CG, Lobarinas E, Popper AN, Fay RR (eds) Translational research in audiology and the hearing sciences, Springer handbook of auditory research. Springer, New York, pp 241–265

114. Plomp R (1986) A signal-to-noise ratio model for the speech-reception threshold of the hearing impaired. J Speech Hear Res 29(2):146–154

115. Wilson RH, McArdle R (2007) Intra- and inter-session test, retest reliability of the Words-in-Noise (WIN) test. J Am Acad Audiol 18(10):813–825

116. Wilson RH, McArdle RA, Smith SL (2007) An evaluation of the BKB-SIN, HINT, QuickSIN, and WIN materials on listeners with normal hearing and listeners with hearing loss. J Speech Lang Hear Res 50(4):844–856. https://doi.org/10.1044/1092-4388(2007/059)

117. Maggirwar SB, Dhanraj DN, Somani SM, Ramkumar V (1994) Adenosine acts as an endogenous activator of the cellular antioxidant defense system. Biochem Biophys Res Commun 201(2):508–515. https://doi.org/10.1006/bbrc.1994.1731

118. Hu BH, Zheng XY, McFadden SL, Kopke RD, Henderson D (1997) R-phenylisopropyladenosine attenuates noise-induced hearing loss in the chinchilla. Hear Res 113(1–2):198–206

119. Hight NG, McFadden SL, Henderson D, Burkard RF, Nicotera T (2003) Noise-induced hearing loss in chinchillas pre-treated with glutathione monoethylester and R-PIA. Hear Res 179(1–2):21–32

120. Samson J, Wiktorek-Smagur A, Politanski P, Rajkowska E, Pawlaczyk-Luszczynska M, Dudarewicz A et al (2008) Noise-induced time-dependent changes in oxidative stress in the mouse cochlea and attenuation by D-methionine. Neuroscience 152(1):146–150. https://doi.org/10.1016/j.neuroscience.2007.11.015

121. Kopke RD, Coleman JK, Liu J, Campbell KC, Riffenburgh RH (2002) Candidate's thesis: enhancing intrinsic cochlear stress defenses to reduce noise-induced hearing loss. Laryngoscope 112(9):1515–1532. https://doi.org/10.1097/00005537-200209000-00001

122. Campbell KC, Meech RP, Klemens JJ, Gerberi MT, Dyrstad SS, Larsen DL et al (2007) Prevention of noise- and drug-induced hearing loss with D-methionine. Hear Res 226(1–2):92–103. https://doi.org/10.1016/j.heares.2006.11.012

123. Campbell K, Claussen A, Meech R, Verhulst S, Fox D, Hughes L (2011) D-methionine (D-met) significantly rescues noise-induced hearing loss: timing studies. Hear Res 282(1–2):138–144. https://doi.org/10.1016/j.heares.2011.08.003

124. Lo WC, Liao LJ, Wang CT, Young YH, Chang YL, Cheng PW (2013) Dose-dependent effects of D-methionine for rescuing noise-induced permanent threshold shift in guinea-pigs. Neuroscience 254:222–229. https://doi.org/10.1016/j.neuroscience.2013.09.027

125. Eastwood H, Pinder D, James D, Chang A, Galloway S, Richardson R, O'Leary S (2010) Permanent and transient effects of locally delivered n-acetyl cysteine in a guinea pig model of cochlear implantation. Hear Res 259(1–2):24–30. https://doi.org/10.1016/j.heares.2009.08.010

126. Zou J, Bretlau P, Pyykko I, Toppila E, Olovius NP, Stephanson N et al (2003) Comparison of the protective efficacy of neurotrophins and antioxidants for vibration-induced trauma. ORL J Otorhinolaryngol Relat Spec 65(3):155–161. https://doi.org/10.1159/000072253

127. Kopke RD, Weisskopf PA, Boone JL, Jackson RL, Wester DC, Hoffer ME et al (2000) Reduction of noise-induced hearing loss using L-NAC and salicylate in the chinchilla. Hear Res 149(1–2):138–146

128. Bielefeld EC, Kopke RD, Jackson RL, Coleman JK, Liu J, Henderson D (2007) Noise protection with N-acetyl-l-cysteine (NAC) using a variety of noise exposures, NAC doses, and routes of administration. Acta Otolaryngol 127(9):914–919. https://doi.org/10.1080/00016480601110188

129. Duan M, Qiu J, Laurell G, Olofsson A, Counter SA, Borg E (2004) Dose and time-dependent protection of the antioxidant N-L-acetylcysteine against impulse noise trauma. Hear Res 192(1–2):1–9. https://doi.org/10.1016/j.heares.2004.02.005

130. Kopke R, Bielefeld E, Liu J, Zheng J, Jackson R, Henderson D, Coleman JK (2005) Prevention of impulse noise-induced hearing loss with antioxidants. Acta Otolaryngol 125(3):235–243

131. Choi CH, Du X, Floyd RA, Kopke RD (2014) Therapeutic effects of orally administrated antioxidant drugs on acute noise-induced hearing loss. Free Radic Res 48(3):264–272. https://doi.org/10.3109/10715762.2013.861599

132. Clifford RE, Coleman JK, Balough BJ, Liu J, Kopke RD, Jackson RL (2011) Low-dose D-methionine and N-acetyl-L-cysteine for protection from permanent noise-induced hearing loss in chinchillas. Otolaryngol Head Neck Surg 145(6):999–1006. https://doi.org/10.1177/0194599811414496

133. Hamernik RP, Qiu W, Davis B (2008) The effectiveness of N-acetyl-L-cysteine (L-NAC) in the prevention of severe noise-induced hearing loss. Hear Res 239(1–2):99–106. https://doi.org/10.1016/j.heares.2008.02.001

134. Lin CY, Wu JL, Shih TS, Tsai PJ, Sun YM, Ma MC, Guo YL (2010) N-acetyl-cysteine against noise-induced temporary threshold shift in male workers. Hear Res 269(1–2):42–47. https://doi.org/10.1016/j.heares.2010.07.005

135. Lindblad AC, Rosenhall U, Olofsson A, Hagerman B (2011) The efficacy of N-acetylcysteine to protect the human cochlea from subclinical hearing loss caused by impulse noise: a controlled trial. Noise Health 13(55):392–401. https://doi.org/10.4103/1463-1741.90293

136. Doosti A, Lotfi Y, Moossavi A, Bakhshi E, Talasaz AH, Hoorzad A (2014) Comparison of the effects of N-acetyl-cysteine and ginseng in prevention of noise induced hearing loss in male textile workers. Noise Health 16(71):223–227. https://doi.org/10.4103/1463-1741.137057

137. Kopke R, Slade MD, Jackson R, Hammill T, Fausti S, Lonsbury-Martin B et al (2015) Efficacy and safety of N-acetylcysteine in prevention of noise induced hearing loss: a randomized clinical trial. Hear Res 323:40–50. https://doi.org/10.1016/j.heares.2015.01.002

138. Kil J, Pierce C, Tran H, Gu R, Lynch ED (2007) Ebselen treatment reduces noise induced hearing loss via the mimicry and induction of glutathione peroxidase. Hear Res 226(1–2):44–51. https://doi.org/10.1016/j.heares.2006.08.006

139. Pourbakht A, Yamasoba T (2003) Ebselen attenuates cochlear damage caused by acoustic trauma. Hear Res 181(1–2):100–108

140. Lynch ED, Gu R, Pierce C, Kil J (2004) Ebselen-mediated protection from single and repeated noise exposure in rat. Laryngoscope 114(2):333–337. https://doi.org/10.1097/00005537-200402000-00029

141. Yamasoba T, Pourbakht A, Sakamoto T, Suzuki M (2005) Ebselen prevents noise-induced excitotoxicity and temporary threshold shift. Neurosci Lett 380(3):234–238. https://doi.org/10.1016/j.neulet.2005.01.047

142. Kil J, Lobarinas E, Spankovich C, Griffiths SK, Antonelli PJ, Lynch ED, Le Prell CG (2017) Safety and efficacy of ebselen for the prevention of noise-induced hearing loss: a randomised, double-blind, placebo-controlled, phase 2 trial. Lancet 390(10098):969–979. https://doi.org/10.1016/S0140-6736(17)31791-9

143. Traber MG, Atkinson J (2007) Vitamin E, antioxidant and nothing more. Free Radic Biol Med 43(1):4–15. https://doi.org/10.1016/j.freeradbiomed.2007.03.024

144. Yamashita D, Jiang HY, Le Prell CG, Schacht J, Miller JM (2005) Post-exposure treatment attenuates noise-induced hearing loss. Neuroscience 134(2):633–642. https://doi.org/10.1016/j.neuroscience.2005.04.015

145. Le Prell CG, Gagnon PM, Bennett DC, Ohlemiller KK (2011) Nutrient-enhanced diet reduces noise-induced damage to the inner ear and hearing loss. Transl Res 158(1):38–53. https://doi.org/10.1016/j.trsl.2011.02.006

146. Chen L, Dean C, Gandolfi M, Nahm E, Mattiace L, Kim AH (2014) Dexamethasone's effect in the retrocochlear auditory centers of a noise-induced hearing loss mouse model. Otolaryngol Head Neck Surg 151(4):667–674. https://doi.org/10.1177/0194599814545771

147. Tabuchi K, Murashita H, Tobita T, Oikawa K, Tsuji S, Uemaetomari I, Hara A (2005) Dehydroepiandrosterone sulfate reduces acoustic injury of the guinea-pig cochlea. J Pharmacol Sci 99(2):191–194

148. Takemura K, Komeda M, Yagi M, Himeno C, Izumikawa M, Doi T et al (2004) Direct inner ear infusion of dexamethasone attenuates noise-induced trauma in guinea pig. Hear Res 196(1–2):58–68. https://doi.org/10.1016/j.heares.2004.06.003

149. Han MA, Back SA, Kim HL, Park SY, Yeo SW, Park SN (2015) Therapeutic effect of dexamethasone for noise-induced hearing loss: systemic versus intratympanic injection in mice. Otol Neurotol 36(5):755–762. https://doi.org/10.1097/MAO.0000000000000759

150. Tabuchi K, Murashita H, Sakai S, Hoshino T, Uemaetomari I, Hara A (2006) Therapeutic time window of methylprednisolone in acoustic injury. Otol Neurotol 27(8):1176–1179. https://doi.org/10.1097/01.mao.0000226313.82069.3f

151. Bas E, Martinez-Soriano F, Lainez JM, Marco J (2009) An experimental comparative study of dexamethasone, melatonin and tacrolimus in noise-induced hearing loss. Acta Otolaryngol 129(4):385–389. https://doi.org/10.1080/00016480802566279

152. Aragno M, Parola S, Brignardello E, Mauro A, Tamagno E, Manti R et al (2000) Dehydroepiandrosterone prevents oxidative injury induced by transient ischemia/reperfusion in the brain of diabetic rats. Diabetes 49(11):1924–1931

153. Kimonides VG, Khatibi NH, Svendsen CN, Sofroniew MV, Herbert J (1998) Dehydroepiandrosterone (DHEA) and DHEA-sulfate (DHEAS) protect hippocampal neurons against excitatory amino acid-induced neurotoxicity. Proc Natl Acad Sci U S A 95(4):1852–1857

154. Morin C, Zini R, Simon N, Tillement JP (2002) Dehydroepiandrosterone and alpha-estradiol limit the functional alterations of rat brain mitochondria submitted to different experimental stresses. Neuroscience 115(2):415–424. https://doi.org/10.1016/s0306-4522(02)00416-5
155. Psillas G, Pavlidis P, Karvelis I, Kekes G, Vital V, Constantinidis J (2008) Potential efficacy of early treatment of acute acoustic trauma with steroids and piracetam after gunshot noise. Eur Arch Otorhinolaryngol 265(12):1465–1469. https://doi.org/10.1007/s00405-008-0689-6
156. Harada H, Shiraishi K, Kato T (2001) Prognosis of acute acoustic trauma: a retrospective study using multiple logistic regression analysis. Auris Nasus Larynx 28(2):117–120
157. Wada T, Sano H, Nishio SY, Kitoh R, Ikezono T, Iwasaki S et al (2017) Differences between acoustic trauma and other types of acute noise-induced hearing loss in terms of treatment and hearing prognosis. Acta Otolaryngol 137(Suppl 565):S48–S52. https://doi.org/10.1080/0001 6489.2017.1297899
158. Himmelseher S, Durieux ME (2005) Revising a dogma: ketamine for patients with neurological injury? Anesth Analg 101(2):524–534, table of contents. https://doi.org/10.1213/01. ANE.0000160585.43587.5B
159. Proescholdt M, Heimann A, Kempski O (2001) Neuroprotection of S(+) ketamine isomer in global forebrain ischemia. Brain Res 904(2):245–251
160. Reeker W, Werner C, Mollenberg O, Mielke L, Kochs E (2000) High-dose S(+)-ketamine improves neurological outcome following incomplete cerebral ischemia in rats. Can J Anaesth 47(6):572–578. https://doi.org/10.1007/BF03018950
161. Himmelseher S, Pfenninger E, Georgieff M (1996) The effects of ketamine-isomers on neuronal injury and regeneration in rat hippocampal neurons. Anesth Analg 83(3):505–512
162. Church J, Zeman S, Lodge D (1988) The neuroprotective action of ketamine and MK-801 after transient cerebral ischemia in rats. Anesthesiology 69(5):702–709
163. Hoffman WE, Pelligrino D, Werner C, Kochs E, Albrecht RF, Schulte am Esch J (1992) Ketamine decreases plasma catecholamines and improves outcome from incomplete cerebral ischemia in rats. Anesthesiology 76(5):755–762
164. Shapira Y, Lam AM, Eng CC, Laohaprasit V, Michel M (1994) Therapeutic time window and dose response of the beneficial effects of ketamine in experimental head injury. Stroke 25(8):1637–1643
165. Bing D, Lee SC, Campanelli D, Xiong H, Matsumoto M, Panford-Walsh R et al (2015) Cochlear NMDA receptors as a therapeutic target of noise-induced tinnitus. Cell Physiol Biochem 35(5):1905–1923. https://doi.org/10.1159/000374000
166. Engelhard K, Werner C, Mollenberg O, Kochs E (2001) S(+)-ketamine/propofol maintain dynamic cerebrovascular autoregulation in humans. Can J Anaesth 48(10):1034–1039. https://doi.org/10.1007/BF03016597
167. Langsjo JW, Maksimow A, Salmi E, Kaisti K, Aalto S, Oikonen V et al (2005) S-ketamine anesthesia increases cerebral blood flow in excess of the metabolic needs in humans. Anesthesiology 103(2):258–268
168. Puel JL, Saffiedine S, Gervais d'Aldin C, Eybalin M, Pujol R (1995) Synaptic regeneration and functional recovery after excitotoxic injury in the guinea pig cochlea. C R Acad Sci III 318(1):67–75
169. Ruel J, Chabbert C, Nouvian R, Bendris R, Eybalin M, Leger CL et al (2008) Salicylate enables cochlear arachidonic-acid-sensitive NMDA receptor responses. J Neurosci 28(29):7313–7323. https://doi.org/10.1523/JNEUROSCI.5335-07.2008
170. Eshraghi AA, Gupta C, Van De Water TR, Bohorquez JE, Garnham C, Bas E, Talamo VM (2013) Molecular mechanisms involved in cochlear implantation trauma and the protection of hearing and auditory sensory cells by inhibition of c-Jun-N-terminal kinase signaling. Laryngoscope 123(Suppl 1):S1–S14. https://doi.org/10.1002/lary.23902
171. Scarpidis U, Madnani D, Shoemaker C, Fletcher CH, Kojima K, Eshraghi AA et al (2003) Arrest of apoptosis in auditory neurons: implications for sensorineural preservation in cochlear implantation. Otol Neurotol 24(3):409–417

172. Wang JC, Raybould NP, Luo L, Ryan AF, Cannell MB, Thorne PR, Housley GD (2003) Noise induces up-regulation of P2X2 receptor subunit of ATP-gated ion channels in the rat cochlea. Neuroreport 14(6):817–823. https://doi.org/10.1097/01.wnr.0000067784.69995.47

173. Ylikoski J, Xing-Qun L, Virkkala J, Pirvola U (2002) Blockade of c-Jun N-terminal kinase pathway attenuates gentamicin-induced cochlear and vestibular hair cell death. Hear Res 166(1–2):33–43

174. Pirvola U, Xing-Qun L, Virkkala J, Saarma M, Murakata C, Camoratto AM et al (2000) Rescue of hearing, auditory hair cells, and neurons by CEP-1347/KT7515, an inhibitor of c-Jun N-terminal kinase activation. J Neurosci 20(1):43–50

175. Wang Y, Ji HX, Xing SH, Pei DS, Guan QH (2007) SP600125, a selective JNK inhibitor, protects ischemic renal injury via suppressing the extrinsic pathways of apoptosis. Life Sci 80(22):2067–2075. https://doi.org/10.1016/j.lfs.2007.03.010

176. Basu S, Kolesnick R (1998) Stress signals for apoptosis: ceramide and c-Jun kinase. Oncogene 17(25):3277–3285. https://doi.org/10.1038/sj.onc.1202570

177. Coleman JK, Littlesunday C, Jackson R, Meyer T (2007) AM-111 protects against permanent hearing loss from impulse noise trauma. Hear Res 226(1–2):70–78. https://doi.org/10.1016/j.heares.2006.05.006

178. Playford MP, Schaller MD (2004) The interplay between Src and integrins in normal and tumor biology. Oncogene 23(48):7928–7946. https://doi.org/10.1038/sj.onc.1208080

179. Harris KC, Hu B, Hangauer D, Henderson D (2005) Prevention of noise-induced hearing loss with Src-PTK inhibitors. Hear Res 208(1–2):14–25. https://doi.org/10.1016/j.heares.2005.04.009

180. Frisch SM, Francis H (1994) Disruption of epithelial cell-matrix interactions induces apoptosis. J Cell Biol 124(4):619–626

181. Bielefeld EC, Hangauer D, Henderson D (2011) Protection from impulse noise-induced hearing loss with novel Src-protein tyrosine kinase inhibitors. Neurosci Res 71(4):348–354. https://doi.org/10.1016/j.neures.2011.07.1836

182. Bielefeld EC, Hynes S, Pryznosch D, Liu J, Coleman JK, Henderson D (2005) A comparison of the protective effects of systemic administration of a pro-glutathione drug and a Src-PTK inhibitor against noise-induced hearing loss. Noise Health 7(29):24–30

183. Bielefeld EC, Wantuck R, Henderson D (2011) Postexposure treatment with a Src-PTK inhibitor in combination with N-l-acetyl cysteine to reduce noise-induced hearing loss. Noise Health 13(53):292–298. https://doi.org/10.4103/1463-1741.82962

184. Miller FD, Pozniak CD, Walsh GS (2000) Neuronal life and death: an essential role for the p53 family. Cell Death Differ 7(10):880–888. https://doi.org/10.1038/sj.cdd.4400736

185. Hu BH, Cai Q, Manohar S, Jiang H, Ding D, Coling DE, Zheng G, Salvi R (2009) Differential expression of apoptosis-related genes in the cochlea of noise-exposed rats. Neuroscience 7;161(3):915–925. https://doi.org/10.1016/j.neuroscience.2009.03.072

186. Fetoni AR, Bielefeld EC, Paludetti G, Nicotera T, Henderson D (2014) A putative role of p53 pathway against impulse noise induced damage as demonstrated by protection with pifithrin-alpha and a Src inhibitor. Neurosci Res 81–82:30–37. https://doi.org/10.1016/j.neures.2014.01.006

Review of Ototoxic Drugs and Treatment Strategies for Reducing Hearing Loss

Chaitanya Mamillapalli, Asmita Dhukhwa, Sandeep Sheth,
Debashree Mukherjea, Leonard P. Rybak, and Vickram Ramkumar

1 Significance and Background

Hearing is one of the essential senses which allows us to perceive sound from the environment. It is important for language and speech development as well as cognitive development. Thus, loss of hearing can adversely affect a person's quality of life. A 2018 World Health Organization (WHO) estimate indicated that more than 5% of world's population suffers from hearing loss. Children are the most vulnerable group in the population, as hearing loss can not only delay language and speech development but also lead to some cognitive disability. Untreated, hearing loss is associated with depression, lower productivity on the job, and social isolation.

C. Mamillapalli
Division of Endocrinology, Department of Internal Medicine, Southern Illinois University School of Medicine, Springfield, IL, USA

Department of Endocrinology, Springfield, Clinic, Springfield, IL, USA

A. Dhukhwa · V. Ramkumar (✉)
Department of Pharmacology, Southern Illinois University School of Medicine, Springfield, IL, USA
e-mail: vramkumar@siumed.edu

S. Sheth
Department of Pharmaceutical Sciences, Larkin University College of Pharmacy, Miami, FL, USA

D. Mukherjea
Division of Otolaryngology, Department of Surgery, Southern Illinois University School of Medicine, Springfield, IL, USA

L. P. Rybak
Department of Pharmacology, Southern Illinois University School of Medicine, Springfield, IL, USA

Division of Otolaryngology, Department of Surgery, Southern Illinois University School of Medicine, Springfield, IL, USA

Hearing loss is categorized into three main categories, these being conductive hearing loss, sensorineural hearing loss, and mixed hearing loss. Of these, sensorineural hearing loss (SNHL) is the most common type, accounting for almost 90% of reported hearing loss [1]. SNHL is characterized by a loss of cells in the sensory epithelium cells in cochlea or damage to the auditory nerve pathways. Multifactorial etiologies for SNHL include prolonged exposure to noise, ototoxic drugs, infection, autoimmune disease, and presbycusis. Loss of sensory cells in the human cochlea leads to permanent hearing loss because of the inability of these cells to regenerate [2, 3]. This review will focus on the most common drugs known to cause hearing loss in human, namely cisplatin and platinum-based chemotherapeutic drugs and aminoglycosides. Some mention will also be made of cyclodextrins, which are used as carriers for lipophilic drugs and produce hearing loss.

2   Cisplatin Ototoxicity

Cisplatin, cis-diamminedichloro platinum [4], was the first FDA-approved platinum drug used for cancer treatment in 1978 [5]. Since then, it has been a well-known chemotherapeutic agent used for treatment of solid tumors such as head and neck, bladder, lung, ovarian, and testicular cancers [6]. It is a platinum coordination compound with a square planar geometry which shows antitumor efficacy by forming covalent bond between platinum atom and DNA purine bases in the N7 position [7]. Despite its wide use over decades, cisplatin possesses significant dose-limiting side effects such as ototoxicity, neurotoxicity, and nephrotoxicity [8]. While nephrotoxicity can be managed by increasing hydration as well as diuresis, there are no FDA-approved treatments available for neurotoxicity and ototoxicity.

Ototoxicity involves cellular degeneration of cochlear tissues that compromises one's functional ability to perceive sound [9]. Cisplatin-induced ototoxicity affects almost 60% of children and up to 23–50% of adults [10, 11]. This drug produces dose-limiting, bilateral, progressive, and irreversible hearing loss. Some studies have reported that almost 100% of cisplatin-treated cancer patients experience increases in hearing thresholds [12, 13]. Hearing loss, especially occurring at an early age, could adversely affect cognitive development and the quality of life of cancer patients [14–16].

2.1   Entry of Cisplatin into the Cochlea

To understand cisplatin ototoxicity, it is important to understand how it enters into the cochlea and why this organ is especially susceptible to this drug. Cisplatin accumulates in hair cells, supporting cells, spiral ganglion cells, and cells in the stria vascularis [17]. One of the well-established mechanism of entry of cisplatin is via copper transporter 1 (CTR1) [18, 19]. Cisplatin binds to the extracellular methionine-rich domain of CTR1, which are highly expressed in the cochlea, thereby facilitating its entry into the cell [20]. Intratympanic administration of copper sulfate (a CTR1 substrate) protects against cisplatin-induced hearing loss [19] by competing with cisplatin for entry into the cell. Cisplatin also enters via cell membrane proteins such as organic cation transporters (OCT1-3) (or SLC22A1-3) [21]. OCT2 is expressed in outer and inner hair cells (IHCs) and stria vascularis (but not spiral ganglion neurons) in the cochlea [22]. OCT2 expression in the stria vascularis was greater in the basal turn. Inhibition or knockout of these transporters protects against cisplatin toxicity [19, 23, 24]. Cisplatin also gains entry into cells by passive diffusion and facilitated transport [25, 26]. Mechanoelectrical transduction [27] channels could represent another major target of entry of cisplatin into the hair cells. Cisplatin has previously been shown to block MET channel in a dose- and voltage-dependent manner in cochlear hair cells derived from chicks [28]. Chemical inhibition of MET using quinine or EGTA prevented cisplatin uptake and protected against cisplatin-induced hair cell death [29]. Another class of nonselective cationic channels, such as TRPV1, TRPV4, TRPA1, TRPC3, and TRPML3, are expressed in cochlea [30, 31] and could serve as additional points of entry of cisplatin into cochlear cells.

Cisplatin exposure has been shown to increase the expression of TRPA1 and TRPV1 in the mouse trigeminal ganglion [32]. The uptake of gentamicin-tagged Texas red (GTTR) via TRPA1 channels has been demonstrated [33], which suggests that the entry of cisplatin (a smaller molecule) into cells could also occur through TRPA1 [33, 34]. TRPV1 expression is also increased in the cochlea after cisplatin treatment. Increased expression is observed in almost all cell types in the cochlea including sensory hair cells, supporting cells, SGNs, and stria vascularis [35]. Moreover, knockdown of TRPV1 using siRNA protected against cisplatin-induced hearing loss in rats [36]. L type and T type Ca^{2+} channels are also present in cochlea [37–39] and their inhibition decreases cisplatin-induced toxicity [40, 41]. This raises the possibility of interaction between cisplatin and these Ca^{2+} channels in vivo. Additionally, the presence of chloride channels such as CFTR and CLC3 and other multidrug proteins in the cochlea could permit cisplatin entry into the cochlea [42–44].

2.2 Mechanisms Underlying Cisplatin Ototoxicity

Increased Oxidative Stress

Cisplatin ototoxicity is closely linked with increased oxidative stress in the cochlea (Fig. 1) [45, 46] and involves either stimulating reactive oxygen species (ROS) generating enzymes or by inactivating antioxidant defense systems [47, 48]. For example, cisplatin reduced cochlear glutathione and antioxidant enzyme activities [49]. A primary target of cisplatin for generation of ROS is the NOX3 NADPH oxidase system [50]. NOX3 is induced by cisplatin and knockdown of this enzyme by transtympanic delivery of siRNA protected against cisplatin-induced ototoxicity [51].

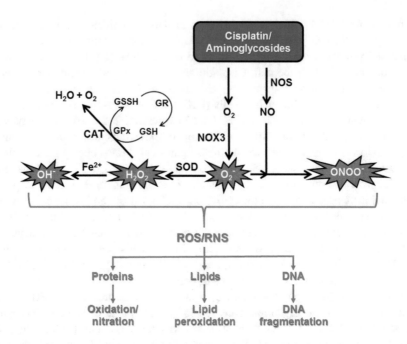

Fig. 1 Schematic of cochlear oxidative stress pathway induced by ototoxic drugs. Reactive oxygen species (ROS) are produced when ototoxic drugs, such as cisplatin and aminoglycosides, react with molecular oxygen (O_2) to generate relatively inert superoxide radical (O_2^-) with the help of NAPDH oxidase-3 (NOX3). Dismutation of O_2^- by superoxide dismutase (SOD) produces a relatively more reactive molecule, hydrogen peroxide (H_2O_2). In the presence of metal ions, such as Fe^{2+}, H_2O_2 is converted into the highly reactive and damaging hydroxyl radical (OH^-). Alternatively, catalase (CAT) can detoxify H_2O_2 to form water (H_2O) and oxygen. Glutathione peroxidase (GPx) can also convert H_2O_2 to H_2O by consuming reduced glutathione (GSH) and producing oxidized glutathione (GSSG) in the process. Glutathione reductase (GR) can reduce GSSG back to GSH. Reactive nitrogen species (RNS) result from the generation of nitric oxide (NO) by nitric oxide synthase (Choe, Chinosornvatana, and Chang). Reaction of NO with O_2^- further generates highly reactive peroxynitrite ($ONOO^-$). ROS and RNS can react with and damage proteins, lipids, and DNA within the cell, thus inhibiting their normal function

Since NOX3 is localized primarily in the cochlea, systemic administration of pharmacological inhibitors [52, 53] could effectively reduce enzyme activity and treat hearing loss. Cisplatin ototoxicity results from the culmination of various oxidative stress and apoptotic processes which occur in the cochlea and overwhelm the endogenous protective mechanisms of the cochlea, such as the antioxidant defense system. Cisplatin increases oxidative stress which derives, in large part, from NOX3 [50]. NOX3 mRNA is highly expressed in the vestibular and cochlear sensory epithelia and in spiral ganglion neurons. Interestingly, NOX3 possess a low level of activity when expressed alone in human embryonic kidney 293 cells; its activity is enhanced when co-expressed with cytoplasmic NOX subunits, p47[phox] and p67[phox]. NOX3 activity is strongly stimulated by the activator protein NOXO1 in the absence of NOXA1 or p47[phox] [54]. Phosphorylation of p47[phox] by protein kinase C (PKC) is essential for its translocation to the membrane and serves as an alternate means of activating NOX3. Since cisplatin can activate PKC, it is reasonable to hypothesize that cisplatin activation of NOX3 might be mediated via activation of PKC. In addition to being activated by cisplatin, a number of studies have also shown that NOX3 expression could be induced by cisplatin through the generation of ROS. This could extend the duration of the oxidative stress in the cochlea initiated by cisplatin.

Another potential source of ROS generation in the cochlea is from xanthine oxidase [55]. This enzyme converts hypoxanthine (a metabolite derived from the breakdown of adenosine by adenosine deaminase) to uric acid. Inhibition of this enzyme by allopurinol contributes to reductions in cisplatin-induced ototoxicity and nephrotoxicity when administered with ebselen [55]. In addition, an earlier report [56] showed that allopurinol alleviated cochlear dysfunction caused by ischemia–perfusion injury in the cochlea. However, a recent study disputed this contention by showing that allopurinol did not improve the auditory deficits in the ischemia–perfusion model [57].

Cisplatin can also increase the production of nitric oxide (NO) via the activation of nitric oxide synthases in the cochlea (Fig. 1) [58, 59]. NO can produce nitration of cochlear proteins, leading to alterations in their functions. The level of one such protein, LMO4, is reduced by nitration [60]. Since inhibition of LMO4 nitration protected against cisplatin-induced hearing loss, these data implicate LMO4 as a critical target for cisplatin ototoxicity. One important target of LMO4 is the signal transducer and activator of transcription 3 (STAT3) [61] which confers pro-survival signals [62].

ROS generated in the cochlea are scavenged by an efficient antioxidant defense system. This comprises antioxidants such as vitamin C, vitamin E, and glutathione. In the guinea pig cochlea, the highest levels of glutathione are present in the basal and intermediate cells of the stria vascularis and in cells of the spiral ligament [63] and correlate well with the distribution of glutathione S-transferases [64]. The cochlea also expresses several antioxidant enzymes, including superoxide dismutase (SOD), glutathione peroxidase (GSH.Px), and catalase (CAT). The sequential actions of SOD and catalase lead to the detoxification of O_2^- to O_2 and H_2O. The enzymes glutathione peroxidase and glutathione reductase participate in the regeneration of glutathione from oxidized glutathione (Fig. 1). Both a cytosolic Cu/Zn isoform and a mitochondrial Mn-regulated isoform (Mn-SOD) of SOD are expressed in the cochlea [65]. The latter is localized to metabolically active sites in the cochlea such as stria vascularis (SV), spiral ligament (SL), spiral prominence, spiral limbus, and organ of Corti (OC) [66]. High levels of GSH and SOD were found in stria vascularis and spiral ligament suggesting an important role of the lateral wall and its antioxidant defense in mitigating oxidative stress in the cochlea [63, 67]. ROS detoxification is essential to the cochlea since, if left unchecked, ROS can produce cellular damage evidenced by increasing lipid peroxides, malondialdehyde, and 4-hydroxynonenal (4-HNE).

Apart from antioxidant enzymes, kidney injury molecule-1 (KIM-1) is induced by cochlear stress where it likely promotes repair of damaged cochlear tissue [68].

Several heat shock proteins (HSP-70, HSP-27, HSP-90) have exhibited protective effect against cisplatin-induced hair cell death [69]. In addition to inhibiting apoptotic signaling pathways, HSPs were also found to activate SOD [4]. Kim et al. have documented the role of heme oxygenase-1 (HO-1) in reducing production of ROS and thus protecting against cisplatin-induced apoptosis of auditory cells [70]. Moreover, overexpression of a transcription factor, Nrf2, protects cisplatin toxicity in auditory cells via PI3 kinase–AKT signaling that promotes HO-1 generation [71]. These endogenous molecules exert cytoprotective effects in the cochlea against damage but are overwhelmed by excessive ROS levels which contribute to apoptosis.

Promoting Cellular Apoptosis

Upon entry into the cells, cisplatin is transformed into its more reactive form which can bind various macromolecules including DNA, RNA, membrane phospholipids, and proteins [72, 73]. The resulting action is activation of molecular and cellular pathways promoting cell necrosis and apoptosis. Once inside the cell, cisplatin produces apoptosis primarily by two different mechanisms, namely via p53 and caspases [74–76] and those mediated by protein kinases [77]. Activation of caspase 3 and 9 has been shown in cochlear hair cell lines [78, 79]. This has also been reported in cisplatin-treated cochlear hair cells of guinea pigs [75] and its inhibition protected against hearing loss. Cisplatin triggers induction of Bax in cells and release of cytochrome c from mitochondria, leading to apoptosis [75, 76]. Our lab has also stressed the importance of the Bax/Bcl-xl ratio and showed increased expression of Bax and decreased expression of Bcl-xl (increased Bax/Bcl ratio) in organ of Corti, spiral ligament, and spiral ganglion cells of rats treated with cisplatin. In addition, cisplatin activated extracellular signal regulated kinase (ERK)1/2, NF-κB, and pro-inflammatory cytokines, such as TNFα, IL-6, and IL-1β [80]. Moreover, treatment of cultured cochlear cells (HE/OC1 cells) with NF-κB inhibitors attenuated cisplatin-induced apoptosis and caspase 3 activation [81]. Cisplatin ototoxicity has also been associated with imbalance in the electrochemical gradient of cochlear endolymph. Type 1 fibrocytes in the spiral ligament are responsible for maintaining this ionic balance via Na^+/K^+-ATPase and any alteration in this ionic equilibrium triggers proapoptotic pathways leading to cell death [82, 83]. NF-κB activation was observed in cisplatin-treated animals in the organ of Corti, spiral ligament, and stria vascularis [80]. Cisplatin-induced expression of cleaved caspase 3 in UB/OC1 cells was inhibited by EGCG treatment via STAT1 inhibition [78]. Based on these findings, activation of caspase is the common pathway for cisplatin-induced apoptosis involving different upstream signaling pathways.

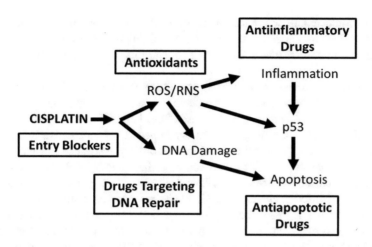

Fig. 2 Potential targets for treating cisplatin ototoxicity. Cisplatin increases oxidative stress in the cochlea which increases cochlear inflammation. Inflammation along with ROS-dependent activation of p53 can promote apoptosis. ROS and RNS can also promote DNA damage (DNA strand breaks). Cisplatin can also modify DNA by interacting with guanine bases, leading to apoptosis. Another potential target of otoprotective agents is blocking the entry of cisplatin into cochlear cells to reduce accumulation of this drug

Covalent Modification of DNA

Cisplatin forms a covalent bond with DNA that interferes with DNA replication and repair system, which ultimately induces apoptosis [84]. Nucleotide excision repair (NER) is involved in repairing damaged DNA, which involves translocation of XPA and XPC into the nucleus [85–87]. Nucleus translocation of both these proteins was observed in the cochlea of cisplatin-treated rats [88]. In addition, increase in ototoxicity was associated with the CC genotype of XPC lys939 Gln [89]. Furthermore, DNA damage and the formation of DNA adducts induce apoptotic signaling via p53, a tumor suppressor gene. The role of p53 has also been studied in cisplatin-mediated auditory cell apoptosis. Pifithrin-α, a p53 inhibitor, prevents hair cell damage caused by cisplatin in organotypic cultures (Fig. 2) [74, 90]. Additionally, p53-deficient mice were resistant to cisplatin-induced apoptosis and showed reduction in caspase 3 activation [74, 91].

2.3 Pharmacogenomics of Cisplatin Ototoxicity

Human studies have identified several genetic variants which are associated with increased sensitivity to cisplatin-induced hearing loss. These genes are involved in the detoxification of cisplatin, reducing oxidative stress and in maintaining intracellular Ca^{2+} homeostasis. Prior knowledge of the genetic polymorphisms inherited by patients could aid in identification of rationale anticancer drug treatments and thereby reduce toxicities. A recent study has highlighted the importance of polymorphisms in several genes linked to susceptibility to cisplatin-induced hearing loss [92].

Glutathione-S-transferases (GSTs) are Phase II enzymes which catalyze the detoxification of xenobiotics by producing a xenobiotic–glutathione conjugate which is more readily excreted. In the case of cisplatin, GSTs conjugation produces a cisplatin–glutathione conjugate which is pumped out of cancer cells by multidrug resistance transporters, enabling development of drug resistance to cisplatin [93]. Three major classes of GSTs have been described, namely GSTP1, GSTT1, and GSTM1, which show differences in their ability to induce glutathione conjugation. Inheritance of different alleles of the GST subclasses has been shown, in some studies, to predict the degree of hearing loss experienced by the patient population. A study by Peters et al. [94], examining patients with a variety of cancers treated with cisplatin, showed that the inheritance of the GSTM3*B allele in human confers protection against cisplatin ototoxicity. Oldenberg et al. [95], in the largest association study of 173 testicular cancer survivor patients, found that the presence of both alleles of 105Val-GSTP1 was protective against cisplatin-induced hearing loss. The same study also reported the detrimental effect of GSTM1 expression on cisplatin-induced hearing loss.

Megalin or the low-density lipoprotein-related protein 2 (LRP2) is a ~600 kD transmembrane protein which serves as a scavenger and Ca^{2+} binding receptor expressed in the human cochlea [96–98]. It binds and transport aminoglycosides into cells of the stria vascularis [99]. It is therefore possible that megalin could also transport cisplatin into the marginal cells. LRP2 polymorphisms, rs2075252 and rs4668123, were positively correlated to cisplatin ototoxicity in a systematic literature review and odds ratio (OR) calculations [92]. LRP2 is also associated with sensorineural deafness and diabetic nephropathy [100].

Acylphosphatase-2 (ACYP-2) is expressed in the cochlea [101] and hydrolyzes the phosphoenzyme intermediate of membrane pumps that affect Ca^{2+} homeostasis [102]. ACYP2 variant, rs1872328, is strongly correlated with cisplatin ototoxicity [103–105]. This suggests a normal role of ACYP-2 in modulating Ca^{2+} homeostasis in the cochlea which is disrupted by cisplatin.

Thiopurine S-methyltransferase (TPMT) has variable enzyme activity depending on polymorphisms. Bhavsar et al. showed that the TPMT*3A variant is associated strongly with cisplatin ototoxicity in vitro and decreased TPMT activity is linked to greater toxicity of anticancer drugs thiopurine and mercaptopurine [106]. Tserga et al. [92] confirmed that TPMT polymorphisms, rs12201199, rs1142345, and rs1800460, are associated with increased cisplatin ototoxicity. Similar conclusions were reached by previous investigators [107–109]. One explanation for cisplatin toxicity associated with TPMT polymorphism is the elevated levels of S-adenosylmethionine in these patients [109].

Catechol-O-methyltransferase (COMT) catalyzes the transfer of a methyl group from S-adenosylmethionine to catecholamines, including the neurotransmitters dopamine, epinephrine, and norepinephrine. COMT is essential for normal physiologic hearing in humans and mice [110]. COMT polymorphism, rs9332377, is an important risk factor in cisplatin ototoxicity in children [109]. These data highlight the importance of catecholamines, such as dopamine, in normal hearing.

The mitochondrial antioxidant enzyme, Mn-SOD2, catalyzes the conversion of O_2^- to H_2O_2. One variant of this enzyme, the rs4880 polymorphism, results in the exchange of valine for alanine. This increases the activity of SOD, resulting in an excessive production of H_2O_2 and hearing loss in the pediatric population treated with cisplatin for medulloblastoma (Brown et al. 2015).

2.4   Approaches to Treating Cisplatin Ototoxicity

Relieving Oxidative Stress

The use of antioxidants is one approach which has been extensively studied for protection against cisplatin-induced hearing loss. Additional approaches such as anti-inflammatory agents are shown in Fig. 2 and are further described below. N-acetylcysteine (NAC) protects against cisplatin-induced hearing loss in rats [111] and guinea pigs [112]. Administration of sodium thiosulfate (STS) protected against cisplatin-induced hearing loss [113]. STS forms a complex with cisplatin which facilitates its inactivation and excretion by the kidneys, thus interfering with the tumoricidal activity of cisplatin. Another antioxidant, amifostine, protects against cisplatin ototoxicity but possesses significant neurotoxicity [114]. D-Methionine, a sulfur-containing amino acid, has also been studied extensively for prevention of cisplatin-induced hearing loss. Both the systemic and local administration of D-methionine effectively reduced cisplatin ototoxicity [115]. D-Methionine also elevated the levels of antioxidant enzymes while reducing the levels of malondialdehyde (marker of lipid peroxidation) after cisplatin administration [116]. A major concern of antioxidant therapy is that the antioxidants could reduce cisplatin chemotherapeutic efficacy [117]. For example, it has been shown that STS reduces the chemotherapeutic efficacy of cisplatin when these drugs were administered simultaneously [118]. Similarly, coadministration of N-acetylcysteine reduced the cytotoxic effects of cisplatin against lung and ovarian cancer cells [119]. Amifostine reduced the levels of cisplatin which is available for treating melanoma without affecting the incidence of ototoxicity [120]. Thus, the utility of these agents could be limited to localized (cochlea) drug delivery to limit interference with cisplatin chemotherapy.

Targeting Cochlear Inflammation

Our laboratory has taken an alternative approach to develop agents which target signaling proteins downstream of ROS with the goal of limiting potential interference with cisplatin chemotherapy. One such target is STAT1 [121–124]. ROS promote Ser^{727} phosphorylation of STAT1, which is crucial for maximal transcriptional activation [125]. STAT1 also mediates cell killing by genotoxic agent such as cisplatin in coordination with the cell cycle regulator, p53 [124]. In fact, cisplatin-mediated activation of STAT1 is absolutely dependent on p53 since it is not observed in p53-null cells [124]. In utricular hair cells, activation of STAT1 promotes cell killing [126]. STAT1 is essential for mediating cisplatin-induced killing of outer hair cells (OHC) in the cochlea, as knockdown of STAT1 by short interfering (si) RNA provided otoprotection [51]. In fact, STAT1 appears to mediate the ototoxicity of other cochlear insults, namely those produced by capsaicin [51], aminoglycoside antibiotics [127], and noise (preliminary data). Thus, small molecule inhibitors of STAT1 can be therapeutically useful for treating ototoxicity.

STAT1 and STAT3 are important regulators of the ischemia/reperfusion and inflammatory process [128]. Downstream gene targets of STAT1 include, but are not limited to, pro-inflammatory genes, *COX-2, iNOS, TNF-α*, vascular cell adhesion molecule (*VCAM*), and intercellular adhesion molecule (*ICAM*). Cisplatin increases the expression of *COX-2, iNOS,* and *TNF-α*. Inhibition of TNF-α by Etanercept reduced cisplatin-induced hearing loss [129]. A previous study by Keithley et al. [130] showed that TNF-α, a master regulator of inflammation, is sufficient to recruit inflammatory cells (leukocytes) to the cochlea. These leukocytes enter the inner ear via the spiral modiolar vein and its tributaries. TNF-α is potentially cytotoxic, acting through the external apoptotic pathway, and thus may participate in the sensory cell degeneration. In contrast, STAT3 protects against oxidative stress [131] and promotes DNA damage repair pathways [132]. Other pro-survival targets of STAT3 include Bcl-xL, Bcl-2, CREB-binding protein (CBP) and caspase-8/MACH/Mch5 (FLICE), FLICE/caspase-8-inhibitory protein (FLIP) and caspase-3 [133]. A recent study from our laboratory revealed exacerbation versus inhibition of cisplatin-induced hearing loss by the STAT1 and STAT3 pathways, respectively [78]. Taken together, these data reinforce the conclusion that targeting STAT1-dependent inflammation/apoptotic pathways and STAT3-dependent survival pathways represents useful otoprotective strategy.

Drugs Targeting G Protein-Coupled Receptors for Treating Cisplatin Ototoxicity

Adenosine A$_1$ receptors (A$_1$AR) are expressed in the organ of Corti, SV, and SGNs [134]. Round window application of the A$_1$AR agonist, *R*-phenylisopropyladenosine (*R*-PIA), in chinchilla increased the activities of antioxidant enzymes, such as GSH. Px and SOD [135]. Moreover, pre-administration of *R*-PIA, significantly reduced cisplatin-induced ABR threshold shifts. Recent studies have further detailed the role of A$_1$AR in otoprotection, highlighting their anti-inflammatory action through inhibition of the ERK/STAT1 pathway [136]. The anti-inflammatory role of the A$_1$AR is also associated with suppression of the NOX3 pathway for ROS generation.

STAT1 is one of the major mediators of cisplatin ototoxicity and inhibition of this transcription factor using EGCG, a STAT1 inhibitor, or by knockdown of STAT1 by siRNA attenuates cisplatin-induced ABR threshold shifts and contributed to otoprotection [51, 78, 126, 136].

Additional studies have documented a protective role of cannabinoid receptor 2 (CB2) agonist against cisplatin-induced apoptosis in auditory cell line [137]. CB2 expression in adult albino rat cochlea was observed predominantly in IHCs, SVs, SG cells, and neurites [138]. Recent study from our laboratory has also elucidated an otoprotective role of CB2 in rat cochlea. Activation of the CB2 receptors by JWH-015 significantly reduced threshold shifts and protected against cisplatin-induced hair cell death and synaptopathy. Importantly, knockdown of CB2 receptors or inhibition by the selective antagonist (AM630) induced hearing loss. This suggests that activation of CB2 receptors by endocannabinoids is essential for maintaining normal hearing.

One study revealed yet another GPCR, Sphingosine-1-phosphate (SIP2) to be a potential target for otoprotection. Activation of SIP2 prevented cisplatin-mediated apoptosis by reducing ROS generation [139].

The importance of GPCRs in various cellular processes including growth and differentiation, neuronal activity, early development, and hormonal release, has been widely studied [140]. Greater understanding of the impact of cochlear-specific GPCR is needed to enable development of targeted drug therapy for hearing loss. One important goal is to regulate the duration of action of these cytoprotective cochlear-specific GPCR. The identification of regulators of G-protein signaling (RGS) proteins whose function is to accelerate the inactivation of active G proteins (G_i, G_o, G_q) will provide a novel therapeutic target for future research.

2.5 Ongoing Clinical Trials for Cisplatin Ototoxicity

Table 1 lists several human studies which have been completed or are ongoing, evaluating the effectiveness of various drugs to treat cisplatin ototoxicity. The trials are listed based on their unique identification codes (NCT number/reference). These include various formulations of dexamethasone for their anti-inflammatory property. Aspirin has also been tested for its anti-inflammatory efficacy for hearing loss. In addition, antioxidants such as alpha lipoic acid, sodium thiosulfate, and Ginkgo biloba extracts have been tested for potential ROS scavengers for cisplatin ototoxicity. Ringer's lactate is believed to protect the cochlea by replenishing lactate for the enzyme lactate dehydrogenase (LDH) and, in particular, the LDH-H isotype, has emerged as markers for cisplatin resistance in multiple tumor types and cell lines [163–165]. LDH is present in abundance in OHC mitochondrial inner membranes and intercristae spaces [166], and perilymph concentrations of LDH and lactate are roughly three times greater than in blood or cerebrospinal fluid [167, 168]. Furthermore, LDH is not inhibited by cisplatin at doses given either clinically or in our study [169]. The LDH-H isoenzyme converts lactate to pyruvate with the gen-

Table 1 Otoprotective agents used in human studies against cisplatin ototoxicity

Otoprotective agent	Mechanism of action	Route of administration	NCT#/ reference	Status
Dexamethasone phosphate	Anti-inflammatory agent; up-regulates cochlear antioxidant enzymes activity	Intratympanic injection	NCT01372904 NCT02382068 [142]	Completed [141] Withdrawn (no patients were enrolled to trial)
OTO-104 (sustained-exposure formulation of dexamethasone)	Anti-inflammatory agent; up-regulates cochlear antioxidant enzymes activity	Intratympanic injection	NCT02997189	Terminated (negative efficacy results from the recently completed Phase 3 study)
Methylprednisolone	Anti-inflammatory agent	Intratympanic injection	NCT01285674	Unknown
Lactated Ringer's solution (with 0.03% Ofloxacin)	Lactate gets converted to pyruvate by lactate dehydrogenase and produces reduced nicotinamide adenine dinucleotide (NADH), an endogenous antioxidant	Administered as ear drops	NCT00584155	Withdrawn (PI left the university)
Ringer's lactate solution (with 0.03% Ciprofloxacin)	Lactate gets converted to pyruvate by lactate dehydrogenase and produces reduced nicotinamide adenine dinucleotide (NADH), an endogenous antioxidant	Transtympanic administration	NCT01108601	Unknown
α-Lipoic acid	Antioxidant	Oral	NCT00477607	Completed
Ginkgo Biloba Extract (GBE761)	Antioxidant	Oral	NCT01139281 [143]	Completed
Aspirin	Anti-inflammatory agent	Oral	NCT00578760	Unknown
Aspirin	Anti-inflammatory agent	Oral	[144]	Completed
Sodium thiosulfate	Antioxidant; binds to and inactivates cisplatin	Transtympanic injection	NCT02281006 [145]	Terminated
Sodium thiosulfate	Antioxidant; binds to and inactivates cisplatin	Intravenous	NCT00716976 NCT00652132 [146–151]	Completed

(continued)

Table 1 (continued)

Otoprotective agent	Mechanism of action	Route of administration	NCT#/ reference	Status
N-Acetylcysteine	Antioxidant; Anti-inflammatory agent	Oral	NCT03400709	Completed
N-Acetylcysteine	Antioxidant; Anti-inflammatory agent	Intravenous administration	NCT02094625	Recruiting
N-Acetylcysteine	Antioxidant; anti-inflammatory agent	Transtympanic injection	[61, 152, 153]	Completed
SPI-1005 (proprietary oral formulation of ebselen)	Mimic and inducer of the enzyme Glutathione peroxidase	Oral	NCT01451853	Not yet recruiting
Amifostine	Antioxidant; inactivates cisplatin	Intravenous	NCT00003269 [154–159]	Completed
Flunarizine	Calcium channel blocker, inhibition of mitochondrial-permeability transition (MPT), suppression of proinflammatory cytokine secretion	Oral	[160]	Completed
Vitamin E	Antioxidant	Oral	[161, 162]	Completed

eration of nicotinamide adenine dinucleotide (NADH), an endogenous antioxidant. Conversely, in a renal organ culture model, cisplatin induces an intracellular NADH depletion that is not reversible with standard thiol-based antioxidants [170].

3 Cyclodextrin-Induced Hearing Loss

Cyclodextrins (CD) are cyclic oligosaccharides produced by starch or starch derivatives. Its chemical structure consists of a hydrophobic cavity formed by glucose units and hydrophilic exterior due to free hydroxyl atoms present outside [171, 172]. These glucopyranose units are linked together which gives cyclodextrins its truncated cone shape. They behave like molecular buckets to entrap lipophilic molecules. Due to this unique structure, cyclodextrins are used in various pharmaceuticals, food, chemical, and cosmetic industrial products to improve solubility, bioavailability, stability, and drug delivery [173–176]. The common CDs consist of 6, 7, or 8 glucopyranose units named as α, β, and γ cyclodextrins, respectively.

Several studies have shown the ability of CDs to manipulate biological lipid compositions such as phospholipids and cholesterol in cells [177, 178]. Among different cyclodextrins, β-cyclodextrins have the highest affinity toward cholesterol.

The synthetic derivatives such as HPβCD and Mβ-CD are widely used β-CDs to modulate cholesterol distribution in cells. Thus, recent studies have emphasized the use of HPβCD not only for its use as excipients but also as an active ingredient in drug development for Niemann Pick Type C disease (NPC) [175, 179–181]. In 2008, Hastings [182] proposed the use of HPβCD for the treatment of NPC disease and the first clinical trial was conducted in humans (C). Based on the results from this study, intrathecal administration of HPβCD for NPC was approved by FDA as an orphan drug for the treatment of NPC [183]. Even though it was FDA approved to be tested for NPC disease, HPβCD was found to be ototoxic for both animals and humans [184–186]. These investigators demonstrated that the doses required to cross the blood–brain barrier to treat neurological diseases such as NPC disease were greater than 4000 mg/kg of HPβCD, doses that resulted in ototoxicity which was not resolved for up to 12 weeks after the cessation of the drug [186]. In this study, repeated administration of 1000 mg/kg of HPβCD had no effect on hearing thresholds, whereas repeated dose of 4000 mg/kg and a single dose of 8000 mg/kg significantly increased hearing thresholds [186]. In addition, Crumling et al. [184] demonstrated the ototoxic effect of a single dose of 8000 mg/kg HPβCD in mice. Both subcutaneous SQ and intracerebroventricular (ICV) delivery of HPβCD elevated hearing thresholds by 40–60 dB [187]. Whole mount preparations suggested that hearing loss was associated with significant loss of OHCs which progressed over a 7-day period from base to apex [187]. Other cochlear cells, including supporting cells, IHCs, and SGNs, were not affected by the drug. The lateral wall suffered limited damage which included edema and vacuolization [187]. Thus, the data suggest that elevations in ABR thresholds induced by HPβCD could result solely from OHC loss. The entry of HPβCD into the inner ear is mediated via the cochlear vasculature (after systemic administration) and via cochlear aqueduct (via central administration) [187, 188]. The reason for the selective OHC toxicity of HPβCD is not clear but could reflect higher levels of cholesterol in OHCs in basal and apical regions of the cochlea compared to the lateral wall [189]. This would suggest that survival and function of OHCs are dependent on their cholesterol content. In this regard, it is estimated that more than 600 proteins in the inner ear are associated with the cholesterol-enriched membrane microdomains [190]. It has been reported that cholesterol can modulate the function of prestin [191, 192] whose function is to carry anionic molecules across the cell membrane [193, 194]. It was only recently shown that OHC-specific HPβCD toxicity was due to prestin. The importance of prestin in OHC function has been attributed to electromotility, maintenance of structural integrity, cochlear amplification, and survival [195–198]. A direct interaction of prestin and cholesterol and OHC survival was observed in prestin knockout mice after HPβCD treatment. HPβCD treatment exacerbated loss of OHC in the presence of prestin whereas OHCs from prestin knockouts were comparatively intact [199]. In addition, cholesterol negatively influences OHC electromotility in a dose-dependent matter via modulation of prestin [200]. Depletion of cholesterol over the development of OHC promotes prestin-associated OHC electromotility, which aid in facilitating graded action potentials to tune hair cell [200, 201]. The cholesterol concentration gradient in the OHC lateral plasma membrane influences prestin-associated charge. Prestin-associated specific charge would be maximal in the region with least cholesterol and cyclodextrin exposure would decrease cholesterol concentration which in turn increases whole cell charge density and cochlear electromechanics [202]. Cholesterol also facilitates prestin trafficking [192, 203]. Thus, this tight relation between cholesterol and prestin could possibly make OHC vulnerable to HPβCD. Zhou et al. [204] showed that the inhibition of prestin's electromotility by salicylate did not ameliorate the ototoxic effect of HPβCD nor did it change HPβCD-induced threshold shifts in both wild type and NPC1 knockout mice. Additionally, 499-prestin knock-in mice which has no motile function phenotypically were also not affected by HPβCD treatment [204]. Therefore, it would be accurate to conclude that sensitivity of OHC toward cyclodextrins is not dependent on electromotility of OHC but rather to the prestin–cholesterol interaction in the lateral membrane of OHC that stabilizes its conformation.

Reactive oxygen species (ROS) are thought to be common pathological mediators for different types of hearing loss such as cisplatin (as described above), aminoglycosides, and noise-induced hearing loss. To determine whether ROS contribute to HPβCD ototoxicity, the efficacy of the ROS scavenger, ebselen, was determined [205]. Ebselen did not protect against HPβCD-induced functional and morphological change to cochlea [205]. A role of cyclodextrins in altering mitochondrial bioenergetics derives from the observation that depletion of cholesterol in mitochondria initiates apoptosis [206]. Beta cyclodextrins such as HPβCD and MβCD produce rapid toxicity toward OHCs at basal turn of the cochlea by a mechanism which is independent of ROS-dependent apoptosis [199, 205, 207]. The acute effects of HPβCD were studied by directly injecting the drug directly into the perilymph of intact cochlea of guinea pig [207]. Exposure of cochlea with high dose of HPβCD elevated cochlear action potential [77], slightly increases the endocochlear potential, decreases DPOAE amplitudes, and abolished auditory nerve overlapped waveform (ANOW) amplitudes [207]. Interestingly, MPβCD showed more drastic damage than HPβCD [207]. It could be due to the ability of MβCD to deplete cholesterol with the highest affinity. This study also demonstrated that neither low nor high dose of HPβCD affected or altered auditory nerve, spiral ganglion neurons, and lateral wall of the scala media.

The increase in HPβCD-induced hearing threshold was observed with no apparent change in central conduction and the ABR waves 1/V amplitudes [186]. This identifies the target site of HPβCD to be within peripheral auditory system. Audiometric and histological findings suggest that the loss of OHCs and its function is the major mechanism behind HPβCD ototoxicity. Cholesterol chelating cyclodextrins could possibly affect OHCs by mobilizing or depleting membrane lipids. The damage to OHCs was more severe in the basal turn of cochlea compared to the apical turn [184, 187, 207]. This observed abrupt transition from damaged to normal hair cells along the length of the cochlea could reflect a differential cholesterol concentration gradient along the cochlea. Depletion of cholesterol by cyclodextrins inhibits K^+ currents and enhances steady state Ca^{2+} current in hair cells [199, 208]. Thus, cyclodextrins can alter hair cell ion homeostasis and ion channel function leading to excitotoxicity and OHC death [184]. It is unclear whether damage extends to other regions of the cochlea or alters synaptic functions. Additional studies are required to determine the mechanisms under-lying cyclodextrin-induced hearing, understanding pharmacokinetics of the drug in the cochlea, unraveling the different molecular pathways involved, and in exploring cell-specific roles of cholesterol in the cochlea.

4 Aminoglycosides

4.1 Overview

Aminoglycosides are broad spectrum antibiotics widely used in clinical practice, with adverse effects to the inner ear and kidney. Aminoglycoside available in the United States includes streptomycin, neomycin, kanamycin, amikacin, gentamicin, tobramycin, plazomycin, and paramomycin. These drugs are potent antimicrobial bactericidal agents with activity against aerobic gram-negative cocci, bacilli, and certain gram-positive cocci. Aminoglycosides are primarily used as a second agent to treat septicemia, intra-abdominal infections, osteomyelitis respiratory tract, and complicated urinary tract infections, caused by aerobic gram-negative bacilli [209]. They are used sometimes in combination with other agents to treat selective gram-positive organisms, such as *Staphylococcus aureus*. Aminoglycosides, such as streptomycin, amikacin, and tobramycin, have antimycobacterial activity and are used as a part of multidrug regimen in the treatment of typical and atypical myco-bacterial infections. The antimicrobial efficacy of aminoglycosides involves bind-ing to 30S ribosome, triggering the misreading of genetic code and premature termination of bacterial protein synthesis. Buildup of the nonfunctional proteins leads to the bacterial death accounting for the bactericidal effects [210, 211].

Ototoxicity was first documented in the first clinical trials of streptomycin in 1946 [212]. Aminoglycoside can cause cochlear toxicity, affecting hearing or ves-tibular toxicity causing imbalance symptoms or both. Streptomycin, gentamycin, tobramycin, and netilmicin are predominantly vestibulotoxic, whereas amikacin, kanamycin, neomycin, and dihydrostreptomycin are primarily cochleotoxic [213]. While vestibulotoxicity can be largely compensated, cochleotoxicity is considered more dangerous.

Sensory hair cells are the mechanoreceptors in the cochlea and are topographi-cally arranged, with hair cells in the basal region tuned primarily to high-frequency sounds, while apical hair cells are tuned largely to low-frequency sounds. Aminoglycoside antibiotics initially damage basal OHCs, stria vascularis, and spi-ral ganglion neurons initially causing high-frequency hearing impairment [214, 215]. Therefore, in the early stages, high-frequency hearing loss (8000–20,000 Hz) goes undetected with a standard clinical audiogram (250–8000 Hz). Hearing loss only becomes clinically manifested after the aminoglycoside lesion progresses toward the low-frequency, apical hair cells.

4.2 Studies Showing Hearing Loss

Aminoglycosides cause hearing loss in up to 2–67% of the patients (see Table 2). This high variation in the incidence of auditory and vestibular aminoglycoside-induced toxicity is attributed to the variation in the methods used to define hearing loss, sole reliance on patient's symptoms, and non-standardization of vestibular testing [228]. The true incidence of ototoxicity is often underestimated as ultrahigh frequencies (>8 Hz) are not routinely tested on traditional hearing tests. In the studies using ultrahigh frequency testing, the incidence of hearing loss with aminoglycoside was high at 67% [220]. Aminoglycosides are retained in the inner ear fluid long after being cleared from the blood stream, and can cause a delayed/progressive ototoxic effect after completion of the treatment. In such situations, hearing loss diagnosed later might not be attributed to the aminoglycoside use [229]. Ototoxicity is not confined to parenteral treatment, but occurs with topical aminoglycoside treatment as well [230]. Several factors contribute to a higher chance of developing ototoxicity. A single high dose of aminoglycoside is more ototoxic than multiple lower doses [231]. In addition, higher doses and longer treatment are associated with higher rates of ototoxicity [232]. Furthermore, coexisting diseases, such as diabetes [233], concurrent administration of other ototoxic drug [234], aging [235], and low dietary intake of proteins [226], could potentiate ototoxicity.

Table 2 Studies demonstrating aminoglycoside ototoxicity [216]

Research study Drug	Research methods	Incidence
Lerner and Matz [217] Gentamicin Amikacin	Prospective study with 54 patients on each drug Pure tone audiometry	Gentamicin 7% Amikacin 9%
Moore et al. [218] Amikacin	Prospective study with 135 patients Pure tone air conduction	22.3%
Fee [219] Gentamicin Tobramycin	Prospective study with 113 patients Pure tone audiometry Hearing loss was defined as a 20 dB neurosensory hearing loss from baseline 33% drop in caloric nystagmus slow phase velocity was used as a marker for vestibular dysfunction	15.3% in the tobramycin group 16.4% gentamicin (Clinically significant toxicity is much lower 4.6% in tobramycin and 2.7% in the gentamicin)
Fausti et al. [220] Aminoglycoside was not specified.	94 EARS High-frequency audiometry The study included both aminoglycosides and cisplatin; ototoxicity outcomes were reported for the combined data	67% using high-frequency ranges 48% using conventional frequency
Hotz et al. [221] Amikacin	10 participants Transient evoked otoacoustic emissions Treatment duration 9–33 days	90%

(continued)

Table 2 (continued)

Research study Drug	Research methods	Incidence
de Jager and van Altena [222] Amikacin Kanamycin Streptomycin	110 participants Pure tone air conduction Mean duration of treatment was 11 weeks of treatment Hearing loss was defined as 15 dB loss at two or more frequencies, or at least 20 dB hearing loss at least one frequency between 0.25 and 8 kHz	18%
Duggal and Sarkar [223] Amikacin Kanamycin Capreomycin	64 participants Pure tone air and bone conduction Mean duration of treatment 18–24 months	18.75% high-frequency loss 6.25% low-frequency loss (Total incidence 25%)
Sturdy et al. [224] Amikacin Capreomycin Streptomycin	Retrospective study Pure tone audiometry Treatment >2 weeks 50 participants Audiogram-based definition ototoxicity • 20 dB loss from baseline at any one test frequency or • 10 dB loss at any two adjacent test frequencies Clinical definition ototoxicity • Hearing loss or tinnitus	28%
Al-Malky et al. [225] Amikacin Tobramycin	Prospective study 39 participants High-frequency pure tone audiometry	21%
Ramma and Ibekwe [226] Kanamycin Amikacin	Prospective study of 53 participants Pure tone audiometry Treatment duration 1–18 months	47%
Harris et al. [227] Capreomycin Kanamycin Streptomycin	Prospective study of 153 participants with MDR-TB Pure tone audiometry	57%

4.3   Uptake of Aminoglycosides into the Cochlea

Aminoglycosides enter hair cells, their primary ototoxic target, through several routes. These include endocytosis of the drug from the endolymph [236, 237] which could partially contribute to hair cell death [236]. Another entry port is megalin, a low-density lipoprotein-related protein 2, which binds aminoglycosides for endocytosis. This protein is highly expressed in the marginal cells of the stria vascularis [99] where it could aid in the clearance of aminoglycosides from the endolymph [238]. Entry can occur through transient receptor potential (TRP) channels, nonselective cation channels, expressed in various cochlear cell types, including spiral ganglion neurons, stria vascularis, and the organ of Corti. Since these channels serve as entry ports for aminoglycosides in sensory neurons [239], it is likely that they also contribute to the entry of aminoglycosides into relevant cochlear cells. Preliminary data from our laboratory support the entry of Texas-Red labeled gentamicin into the organ of Corti via one such TRP channel, the TRPV1 channel (unpublished). TRPV1 is expressed in the organ of Corti and also contributes to cisplatin-induced ototoxicity [51, 240]. Other TRP channels present in the cochlea, including TRPV4, TRPA1, and TRPC3, could also serve as potential uptake sites for aminoglycosides [238].

MET channels are nonspecific cation channels present on the tips of the hair cell stereocilia, which open and close to hair bundle deflection. Mutations in components of the MET channels, such as myosin VIIA [241] and associated proteins, cadherin-23, and protococadherin-15 proteins, reduce the uptake of aminoglycosides into the hair cells [242], leading to increased survival [242]. These data suggest that drugs which block entry should reduce aminoglycoside uptake and reduce its toxicity.

4.4 Aminoglycoside Genetic Susceptibility

The affinity of aminoglycosides for eukaryotic ribosomes is low due to structural difference from non-eukaryotic cells; hence aminoglycoside toxic effects are usually limited to bacteria, making the drug safe for humans. However, certain mitochondrial DNA mutations can increase their affinity of ribosomes to aminoglycosides, thereby increasing their ototoxic effects. The mitochondrial 12S rRNA mutations A1555G mutation and the C1494T mutation create a new G–C or U–A base pair changing the structure of the human mitochondrial 12S rRNA. This renders it more similar to bacterial 16S rRNA, making it a primary target of aminoglycosides binding and increased toxicity [233, 243–245]. In patients with A1555G mutations, hearing loss can occur following a single dose of aminoglycoside and almost all the patients develop hearing impairment following aminoglycoside use [234, 235]. These mitochondrial mutations were reported worldwide and include African, Caucasian, Arab–Israeli, Chinese, Spanish, Japanese, and Mongolian ethnic groups. Prior to treatment, clinicians should obtain detailed family history and consider the possibility of preexisting genetic defects, especially in susceptible racial groups [246]. These mutations have extremely low penetrance, and hence they may not have baseline hearing problems [262]. Recent clinical studies demonstrated the feasibility of targeted pathogenic mutation screening to detect carriers of these mutations, and can possibly help to improve the safety of aminoglycosides [247].

4.5 Preventative Strategies for Ototoxicity

Clinicians must restrict aminoglycosides use for defined clinical indications. Risk-to-benefit ratio should be considered in aminoglycoside treatment and patients should be engaged in the decision-making process. Health care providers must educate patients regarding the potential early symptoms of ototoxicity and encourage patient to report symptoms of hearing loss or imbalance [248]. Patients should avoid loud noise to prevent noise-induced damage during and for at least 6 months post-aminoglycoside treatment [249], as aminoglycosides have a long retention time in the ear. In patients who have higher risk for ototoxicity as discussed above (such as advanced age, renal dysfunction), these agents should be used with caution and alternative agents may be considered. It is important to avoid concomitant use of aminoglycoside with other ototoxic medication, such as loop diuretics and vancomycin, which can potentiate aminoglycoside toxicity [238]. Aminoglycoside drug levels and renal function tests should be carefully monitored to prevent overdosing. Total cumulative exposure of aminoglycoside increases the risk of toxicity; hence, the duration of treatment should be limited to the shortest course as possible.

No universal consensus exists regarding the protocol for ototoxicity monitoring during aminoglycoside treatment. Monitoring program should involve high-frequency audiometry (frequencies above 8 kHz) for the early detection of the hearing impact [214, 250] (see Table 3). With any signs of ototoxicity on the audiogram, aminoglycoside use should be stopped immediately to allow for potential recovery of auditory function. Hearing and vestibular evaluations should be performed in all patients who are on long duration treatment for more than 14 days. Hearing loss can occur as early as 5 days after the start of aminoglycosides [219], and some experts recommend starting the testing before and 1 week after the start of drug treatment and to continue testing once a week thereafter [221]. Due to the lack of resources and awareness among health care providers, these protocols are unfortunately not widely implemented in clinical practice.

Table 3 Auditory assessment of ototoxicity [231]

Conventional pure tone audiometry. 250–8000 Hz	• Detects and measures the severity of hearing loss in the middle frequency range of hearing. • Patient acts as their own control, changes are compared to the baseline values. • Ultrahigh frequency loss is not detected, hence will miss many cases of early ototoxicity. • This can be done at bedside. • Depends on patient's ability to attend to task and is affected by ambient noise.

(continued)

Table 3 (continued)

High-frequency pure tone audiometry Frequencies >20,000 Hz	• Detects and measures the severity of hearing loss in the high-frequency range of hearing. • Test of choice recommended by American Speech–Language Hearing Association (ASHA) and the American Academy of Audiology (AAA) guidelines [251]. • Patients act as their own control, change compared to the baseline values. • More specialized equipment is required. • Depends on patient's ability to attend to task and is affected by ambient noise.
Otoacoustic emissions	• Assesses outer hair cell function. • Active cooperation from the patient is not required, and can be done at bedside. • Not influenced by ambient noise. • This technique has no widely accepted or validated criteria to define significant ototoxicity [252].
Auditory brainstem response	• Measures the function of VIIIth cranial nerve. • Needs only passive cooperation from the patient. • Interpretation requires the expertise of an audiologist.

Antioxidants

Aminoglycoside entry into the hair cells causes an increase in the formation of ROS or free radicals. Antioxidants prevent lipid peroxidation and cell damage by scavenging ROS [253]. A variety of antioxidants have been shown to protect against aminoglycoside ototoxicity as detailed below.

Several studies were performed in animals to test the efficacy of antioxidants to treat aminoglycoside ototoxicity. Administration of alpha-tocopherol protected against gentamicin-induced loss of OHCs in the cochlear basal turn [253]. In a rat experimental model, oral D-methionine protected against amikacin-induced ototoxicity [251]. In addition, α-lipoic acid attenuated cochlear damage produced by aminoglycoside amikacin in guinea pigs [254]. In a study using Hartley albino guinea pigs, concomitant administration of Q-ter (a soluble formulation of coenzyme Q) decreased the severity of gentamicin-induced auditory impairment [255]. Another study in guinea pigs showed that pretreatment with ginkgo biloba prevented gentamicin-induced morphological and functional damage of the cochlea [256].

Studies in human show that *N*-acetylcysteine (NAC), a mucolytic agent with antioxidant properties, protects against aminoglycoside ototoxic effects. In this study of 53 hemodialysis patients receiving gentamicin treatment, coadministration of NAC with gentamicin was associated with less high-frequency hearing threshold shifts compared to patients receiving only gentamicin. NAC treatment was given for an additional 1 week after stopping gentamicin treatment and the protective effects were sustained for up to 6 weeks posttreatment [257]. In a systematic review including three clinical randomized trials, coadministration of NAC with aminoglycosides reduced the risk of ototoxicity by 80% [252]. In clinical trials, the use of acetylsalicylate (ASA) protected against ototoxic damage without affecting the antimicrobial potency of gentamicin [258]. High or prolonged treatment with ASA can cause transient ototoxicity and also other adverse effects like gastrointestinal bleeding. Thus, more research is required to establish the safe dose of ASA as an otoprotective agent [259].

Inhibition of Apoptosis

Apoptotic enzyme inhibitors such as caspase inhibitors, stress kinase blockers, and heat shock protein inducers can protect against hair cell damage in vitro and in vivo. Apoptosis or programmed cell death is a protective mechanism against uncontrolled cellular proliferation. Therefore, prolonged use of these agents is likely unsafe for human use because of the increased risk of carcinogenesis [260].

Mechanoelectrical Transducer Channel Blockers

Aminoglycosides enter hair cells through the MET channels, and blocking these using various agents has been explored as potential treatments for preventing aminoglycoside ototoxicity. For example, d-tubocurarine, a MET channel blocker, decreased gentamicin intake into rat cochlear hair cells and prevented hair cell death [261]. However, many of the MET channel blockers are toxic to humans and therefore are not appropriate for therapeutic use. Unlike other cell types which are able to clear aminoglycosides from their cytoplasm, hair cells and renal proximal tubular cells are able to retain these drugs for extended periods [239], rendering these targets more susceptible to toxicity. Other cochlear cells, such as supporting cells, spiral ganglion neurons, and cells in the lateral wall [239, 263, 264], can also take up aminoglycosides (presumably by different mechanisms), but aminoglycosides are cleared from these cells more rapidly than hair cells, which retain aminoglycosides for at least 6 months [263].

Matt et al. investigated apramycin (an aminoglycoside used in veterinary medicine) and showed that because of its unique structure it has low affinity for mitochondrial ribosomes in eukaryotes, while still preserving its strong bactericidal activity [265]. This observation provided proof of concept for the development of designer aminoglycosides with the aim of dissociating antimicrobial activity from ototoxicity. N1MS, an aminoglycoside congener, which has reduced penetration into hair cell mechanotransducer channels has 17 times lower ototoxicity with preservation of its antibiotic activity [266]. More research is needed to garner additional insights into the impact of aminoglycoside structure changes to maintain antimicrobial activity while reducing ototoxicity. Several other novel mechanisms of aminoglycoside ototoxicity targeting stria vascularis, ribosomes, calcium signaling, lysosomal degradation, plasma membranes, PIP2, and potassium channels are not discussed in this review described elsewhere [234].

5 Summary

In summary, cisplatin and aminoglycosides are important chemotherapeutic agents for treating cancers and infections, respectively. However, the effective use of these agents is limited by significant toxicities, including hearing loss. High-frequency audiometry should be used for early detection of hearing impairment and stopping aminoglycoside treatment at the earliest phase of ototoxicity to reduce the degree of impairment. In most cases, physicians would have to balance the risk versus benefit to the patient in deciding whether to continue drug treatment. Though many compounds have demonstrated otoprotective function, currently none of the medications are approved for prevention of ototoxicity, or to reverse hearing impairment. Hence it is paramount for clinicians to implement preventative strategies that include monitoring or modifying the dose and treatment duration of these drugs. The development of newer platinum congeners and designer antibiotics with limited ototoxic potential could open up a new era for safer treatments of patient populations with these important therapeutic agents.

Acknowledgements The authors would like to acknowledge funding from NIH grants RO1-CA166907 and RO1-DC016835 (to V.R.) and RO1-DC002396 (to L.P.R.), which support studies from the authors' laboratories described in this review.

References

1. Schuknecht HF, Kimura RS, Naufal PM (1973) The pathology of sudden deafness. Acta Otolaryngol 76:75–97
2. Chardin S, Romand R (1995) Regeneration and mammalian auditory hair cells. Science 267:707–711
3. Groves AK (2010) The challenge of hair cell regeneration. Exp Biol Med (Maywood) 235:434–446
4. Moriyama-Gonda N, Igawa M, Shiina H, Urakami S, Shigeno K, Terashima M (2002) Modulation of heat-induced cell death in PC-3 prostate cancer cells by the antioxidant inhibitor diethyldithiocarbamate. BJU Int 90:317–325
5. Martelli L, Di Mario F, Botti P, Ragazzi E, Martelli M, Kelland L (2007) Accumulation, platinum-DNA adduct formation and cytotoxicity of cisplatin, oxaliplatin and satraplatin in sensitive and resistant human osteosarcoma cell lines, characterized by p53 wild-type status. Biochem Pharmacol 74:20–27
6. Abrams TJ, Lee LB, Murray LJ, Pryer NK, Cherrington JM (2003) SU11248 inhibits KIT and platelet-derived growth factor receptor beta in preclinical models of human small cell lung cancer. Mol Cancer Therap 2:471–478
7. Wang D, Lippard SJ (2005) Cellular processing of platinum anticancer drugs. Nat Rev Drug Discov 4:307–320
8. Rybak LP, Whitworth CA, Mukherjea D, Ramkumar V (2007) Mechanisms of cisplatin-induced ototoxicity and prevention. Hear Res 226:157–167
9. Arslan E, Orzan E, Santarelli R (1999) Global problem of drug-induced hearing loss. Ann N Y Acad Sci 884:1–14

10. Coradini PP, Cigana L, Selistre SG, Rosito LS, Brunetto AL (2007) Ototoxicity from cisplatin therapy in childhood cancer. J Pediatr Hematol Oncol 29:355–360
11. Knight KR, Kraemer DF, Neuwelt EA (2005) Ototoxicity in children receiving platinum chemotherapy: underestimating a commonly occurring toxicity that may influence academic and social development. J Clin Oncol 23:8588–8596
12. Bisht M, Bist SS (2011) Ototoxicity: the hidden menace. Indian J Otolaryngol Head Neck Surg 63:255–259
13. McKeage MJ (1995) Comparative adverse effect profiles of platinum drugs. Drug Saf 13:228–244
14. Rybak LP, Mukherjea D, Jajoo S, Ramkumar V (2009) Cisplatin ototoxicity and protection: clinical and experimental studies. Tohoku J Exp Med 219:177–186
15. Sheth S, Mukherjea D, Rybak LP, Ramkumar V (2017) Mechanisms of cisplatin-induced ototoxicity and otoprotection. Front Cell Neurosci 11:338
16. van den Berg JH, Beijnen JH, Balm AJ, Schellens JH (2006) Future opportunities in preventing cisplatin induced ototoxicity. Cancer Treat Rev 32:390–397
17. Breglio AM, Rusheen AE, Shide ED, Fernandez KA, Spielbauer KK, McLachlin KM, Hall MD, Amable L, Cunningham LL (2017) Cisplatin is retained in the cochlea indefinitely following chemotherapy. Nat Commun 8:1654
18. Holzer AK, Katano K, Klomp LW, Howell SB (2004) Cisplatin rapidly down-regulates its own influx transporter hCTR1 in cultured human ovarian carcinoma cells. Clin Cancer Res 10:6744–6749
19. More SS, Akil O, Ianculescu AG, Geier EG, Lustig LR, Giacomini KM (2010) Role of the copper transporter, CTR1, in platinum-induced ototoxicity. J Neurosci 30:9500–9509
20. Öhrvik H, Thiele DJ (2015) The role of Ctr1 and Ctr2 in mammalian copper homeostasis and platinum-based chemotherapy. J Trace Elem Med Biol 31:178–182
21. Ciarimboli G, Ludwig T, Lang D, Pavenstadt H, Koepsell H, Piechota HJ, Haier J, Jaehde U, Zisowsky J, Schlatter E (2005) Cisplatin nephrotoxicity is critically mediated via the human organic cation transporter 2. Am J Pathol 167:1477–1484
22. Ciarimboli G, Schlatter E (2005) Regulation of organic cation transport. Pflugers Arch 449:423–441
23. Ciarimboli G, Deuster D, Knief A, Sperling M, Holtkamp M, Edemir B, Pavenstadt H, Lanvers-Kaminsky C, am Zehnhoff-Dinnesen A, Schinkel AH, Koepsell H, Jurgens H, Schlatter E (2010) Organic cation transporter 2 mediates cisplatin-induced oto- and nephrotoxicity and is a target for protective interventions. Am J Pathol 176:1169–1180
24. Sprowl JA, van Doorn L, Hu S, van Gerven L, de Bruijn P, Li L, Gibson AA, Mathijssen RH, Sparreboom A (2013) Conjunctive therapy of cisplatin with the OCT2 inhibitor cimetidine: influence on antitumor efficacy and systemic clearance. Clin Pharmacol Ther 94:585–592
25. Binks SP, Dobrota M (1990) Kinetics and mechanism of uptake of platinum-based pharmaceuticals by the rat small intestine. Biochem Pharmacol 40:1329–1336
26. Burger H, Loos WJ, Eechoute K, Verweij J, Mathijssen RHJ, Wiemer EAC (2011) Drug transporters of platinum-based anticancer agents and their clinical significance. Drug Resist Updat 14:22–34
27. Murai N, Kirkegaard M, Jarlebark L, Risling M, Suneson A, Ulfendahl M (2008) Activation of JNK in the inner ear following impulse noise exposure. J Neurotrauma 25:72–77
28. Kimitsuki T, Ohmori H (1993) Dihydrostreptomycin modifies adaptation and blocks the mechano-electric transducer in chick cochlear hair cells. Brain Res 624:143–150
29. Thomas AJ, Hailey DW, Stawicki TM, Patricia W, Coffin AB, Rubel EW, Raible DW, Simon JA, Henry CO (2013) Functional mechanotransduction is required for cisplatin-induced hair cell death in the zebrafish lateral line. J Neurosci 33:4405–4414
30. Asai Y, Holt JR, Géléoc GSG (2010) A quantitative analysis of the spatiotemporal pattern of transient receptor potential gene expression in the developing mouse cochlea. JARO: J Assoc Res Otolaryngol 11:27–37

31. Cuajungco MP, Grimm C, Heller S (2007) TRP channels as candidates for hearing and balance abnormalities in vertebrates. Biochim Biophys Acta 1772:1022–1027
32. Ta LE, Bieber AJ, Carlton SM, Loprinzi CL, Low PA, Windebank AJ (2010) Transient Receptor Potential Vanilloid 1 is essential for cisplatin-induced heat hyperalgesia in mice. Mol Pain 6:15–15
33. Stepanyan RS, Indzhykulian AA, Velez-Ortega AC, Boger ET, Steyger PS, Friedman TB, Frolenkov GI (2011) TRPA1-mediated accumulation of aminoglycosides in mouse cochlear outer hair cells. J Assoc Res Otolaryngol 12:729–740
34. Lian HY, Hu M, Liu CH, Yamauchi Y, Wu KC (2012) Highly biocompatible, hollow coordination polymer nanoparticles as cisplatin carriers for efficient intracellular drug delivery. Chem Commun (Cambridge, England) 48:5151–5153
35. Phan PA, Tadros SF, Kim Y, Birnbaumer L, Housley GD (2010) Developmental regulation of TRPC3 ion channel expression in the mouse cochlea. Histochem Cell Biol 133:437–448
36. Mukherjea D, Jajoo S, Whitworth C, Bunch JR, Turner JG, Rybak LP, Ramkumar V (2008) Short interfering RNA against transient receptor potential vanilloid 1 attenuates cisplatin-induced hearing loss in the rat. J Neurosci 28:13056–13065
37. Hafidi A, Dulon D (2004) Developmental expression of Ca(v)1.3 (alpha1d) calcium channels in the mouse inner ear. Brain Res Dev Brain Res 150:167–175
38. Lei D, Gao X, Perez P, Ohlemiller KK, Chen C-C, Campbell KP, Hood AY, Bao J (2011) Anti-epileptic drugs delay age-related loss of spiral ganglion neurons via T-type calcium channel. Hear Res 278:106–112
39. Uemaetomari I, Tabuchi K, Nakamagoe M, Tanaka S, Murashita H, Hara A (2009) L-type voltage-gated calcium channel is involved in the pathogenesis of acoustic injury in the cochlea. Tohoku J Exp Med 218:41–47
40. Mohajjel Nayebi A, Sharifi H, Ramadzani M, Rezazadeh H (2012) Effect of acute and chronic administration of carbamazepine on Cisplatin-induced hyperalgesia in rats. Jundishapur J Nat Pharm Prod 7:27–30
41. Muthuraman A, Singla SK, Peters A (2011) Exploring the potential of flunarizine for Cisplatin-induced painful uremic neuropathy in rats. Int Neurourol J 15:127–134
42. Kawasaki E, Hattori N, Miyamoto E, Yamashita T, Inagaki C (1999) Single-cell RT-PCR demonstrates expression of voltage-dependent chloride channels (ClC-1, ClC-2 and ClC-3) in outer hair cells of rat cochlea. Brain Res 838:166–170
43. Oshima T, Ikeda K, Furukawa M, Takasaka T (1997) Expression of voltage-dependent chloride channels in the rat cochlea. Hear Res 103:63–68
44. Saito T, Zhang ZJ, Tokuriki M, Ohtsubo T, Noda I, Shibamori Y, Yamamoto T, Saito H (2001) Expression of multidrug resistance protein 1 (MRP1) in the rat cochlea with special reference to the blood-inner ear barrier. Brain Res 895:253–257
45. Clerici WJ, Hensley K, DiMartino DL, Butterfield DA (1996) Direct detection of ototoxicant-induced reactive oxygen species generation in cochlear explants. Hear Res 98:116–124
46. Kopke R, Staecker H, Lefebvre P, Malgrange B, Moonen G, Ruben RJ, Van de Water TR (1996) Effect of neurotrophic factors on the inner ear: clinical implications. Acta Otolaryngol 116:248–252
47. Church MW, Kaltenbach JA, Blakley BW, Burgio DL (1995) The comparative effects of sodium thiosulfate, diethyldithiocarbamate, fosfomycin and WR-2721 on ameliorating cisplatin-induced ototoxicity. Hear Res 86:195–203
48. Rybak LP, Ravi R, Somani SM (1995) Mechanism of protection by diethyldithiocarbamate against cisplatin ototoxicity: antioxidant system. Fundam Appl Toxicol 26:293–300
49. Ravi R, Somani SM, Rybak LP (1995) Mechanism of cisplatin ototoxicity: antioxidant system. Pharmacol Toxicol 76:386–394
50. Banfi B, Malgrange B, Knisz J, Steger K, Dubois-Dauphin M, Krause KH (2004) NOX3, a superoxide-generating NADPH oxidase of the inner ear. J Biol Chem 279:46065–46072
51. Mukherjea D, Jajoo S, Sheehan K, Kaur T, Sheth S, Bunch J, Perro C, Rybak LP, Ramkumar V (2011) NOX3 NADPH oxidase couples transient receptor potential vanilloid 1 to signal

transducer and activator of transcription 1-mediated inflammation and hearing loss. Antioxid Redox Signal 14:999–1010

52. Cifuentes-Pagano ME, Meijles DN, Pagano PJ (2015) Nox inhibitors & therapies: rational design of peptidic and small molecule inhibitors. Curr Pharm Des 21:6023–6035

53. Rousset F, Carnesecchi S, Senn P, Krause KH (2015) Nox3-targeted therapies for inner ear pathologies. Curr Pharm Des 21:5977–5987

54. Cheng G, Ritsick D, Lambeth JD (2004) Nox3 regulation by NOXO1, p47phox, and p67phox. J Biol Chem 279:34250–34255

55. Lynch ED, Gu R, Pierce C, Kil J (2005) Reduction of acute cisplatin ototoxicity and nephrotoxicity in rats by oral administration of allopurinol and ebselen. Hear Res 201:81–89

56. Seidman MD, Quirk WS, Nuttall AL, Schweitzer VG (1991) The protective effects of allopurinol and superoxide dismutase-polyethylene glycol on ischemic and reperfusion-induced cochlear damage. Otolaryngol Head Neck Surg 105:457–463

57. Tabuchi K, Tsuji S, Ito Z, Hara A, Kusakari J (2001) Does xanthine oxidase contribute to the hydroxyl radical generation in ischemia and reperfusion of the cochlea? Hear Res 153:1–6

58. Li G, Liu W, Frenz D (2006) Cisplatin ototoxicity to the rat inner ear: a role for HMG1 and iNOS. Neurotoxicology 27:22–30

59. Watanabe K, Inai S, Jinnouchi K, Bada S, Hess A, Michel O, Yagi T (2002) Nuclear-factor kappa B (NF-kappa B)-inducible nitric oxide synthase (iNOS/NOS II) pathway damages the stria vascularis in cisplatin-treated mice. Anticancer Res 22:4081–4085

60. Jamesdaniel S, Ding D, Kermany MH, Davidson BA, Knight PR 3rd, Salvi R, Coling DE (2008) Proteomic analysis of the balance between survival and cell death responses in cisplatin-mediated ototoxicity. J Proteome Res 7:3516–3524

61. Riga MG, Chelis L, Kakolyris S, Papadopoulos S, Stathakidou S, Chamalidou E, Xenidis N, Amarantidis K, Dimopoulos P, Danielides V (2013) Transtympanic injections of N-acetylcysteine for the prevention of cisplatin-induced ototoxicity: a feasible method with promising efficacy. Am J Clin Oncol 36:1–6

62. Hilfiker-Kleiner D, Hilfiker A, Fuchs M, Kaminski K, Schaefer A, Schieffer B, Hillmer A, Schmiedl A, Ding Z, Podewski E, Podewski E, Poli V, Schneider MD, Schulz R, Park JK, Wollert KC, Drexler H (2004) Signal transducer and activator of transcription 3 is required for myocardial capillary growth, control of interstitial matrix deposition, and heart protection from ischemic injury. Circ Res 95:187–195

63. Usami S, Hjelle OP, Ottersen OP (1996) Differential cellular distribution of glutathione—an endogenous antioxidant—in the guinea pig inner ear. Brain Res 743:337–340

64. el Barbary A, Altschuler RA, Schacht J (1993) Glutathione S-transferases in the organ of Corti of the rat: enzymatic activity, subunit composition and immunohistochemical localization. Hear Res 71:80–90

65. Yao X, Rarey KE (1996) Detection and regulation of Cu/Zn-SOD and Mn-SOD in rat cochlear tissues. Hear Res 96:199–203

66. Lai MT, Ohmichi T, Egusa K, Okada S, Masuda Y (1996) Immunohistochemical localization of manganese superoxide dismutase in the rat cochlea. Eur Arch Otorhinolaryngol 253:273–277

67. Pierson MG, Gray BH (1982) Superoxide dismutase activity in the cochlea. Hear Res 6:141–151

68. Mukherjea D, Whitworth CA, Nandish S, Dunaway GA, Rybak LP, Ramkumar V (2006) Expression of the kidney injury molecule 1 in the rat cochlea and induction by cisplatin. Neuroscience 139:733–740

69. Cunningham LL, Brandon CS (2006) Heat shock inhibits both aminoglycoside- and cisplatin-induced sensory hair cell death. J Assoc Res Otolaryngol: JARO 7:299–307

70. Kim H-J, So H-S, Lee J-H, Lee J-H, Park C, Park S-Y, Kim Y-H, Youn M-J, Kim S-J, Chung S-Y, Lee K-M, Park R (2006) Heme oxygenase-1 attenuates the cisplatin-induced apoptosis of auditory cells via down-regulation of reactive oxygen species generation. Free Radic Biol Med 40:1810–1819

71. So HS, Kim HJ, Lee JH, Lee JH, Park SY, Park C, Kim YH, Kim JK, Lee KM, Kim KS, Chung SY, Jang WC, Moon SK, Chung HT, Park RK (2006) Flunarizine induces Nrf2-mediated transcriptional activation of heme oxygenase-1 in protection of auditory cells from cisplatin. Cell Death Differ 13:1763–1775

72. Jamieson ER, Lippard SJ (1999) Structure, recognition, and processing of cisplatin-DNA adducts. Chem Rev 99:2467–2498

73. Long DF, Repta AJ (1981) Cisplatin: chemistry, distribution and biotransformation. Biopharm Drug Dispos 2:1–16

74. Benkafadar N, Menardo J, Bourien J, Nouvian R, François F, Decaudin D, Maiorano D, Puel J-L, Wang J (2017) Reversible p53 inhibition prevents cisplatin ototoxicity without blocking chemotherapeutic efficacy. EMBO Mol Med 9:7–26

75. Wang J, Ladrech S, Pujol R, Brabet P, Van De Water TR, Puel JL (2004) Caspase inhibitors, but not c-Jun NH2-terminal kinase inhibitor treatment, prevent cisplatin-induced hearing loss. Cancer Res 64:9217–9224

76. Watanabe K, Inai S, Jinnouchi K, Baba S, Yagi T (2003) Expression of caspase-activated deoxyribonuclease (CAD) and caspase 3 (CPP32) in the cochlea of cisplatin (CDDP)-treated guinea pigs. Auris Nasus Larynx 30:219–225

77. Previati M, Lanzoni I, Astolfi L, Fagioli F, Vecchiati G, Pagnoni A, Martini A, Capitani S (2007) Cisplatin cytotoxicity in organ of corti-derived immortalized cells. J Cell Biochem 101:1185–1197

78. Borse V, Al Aameri RFH, Sheehan K, Sheth S, Kaur T, Mukherjea D, Tupal S, Lowy M, Ghosh S, Dhukhwa A, Bhatta P, Rybak LP, Ramkumar V (2017) Epigallocatechin-3-gallate, a prototypic chemopreventative agent for protection against cisplatin-based ototoxicity. Cell Death Dis 8:e2921

79. Devarajan P, Savoca M, Castaneda MP, Park MS, Esteban-Cruciani N, Kalinec G, Kalinec F (2002) Cisplatin-induced apoptosis in auditory cells: role of death receptor and mitochondrial pathways. Hear Res 174:45–54

80. So H, Kim H, Lee JH, Park C, Kim Y, Kim E, Kim JK, Yun KJ, Lee KM, Lee HY, Moon SK, Lim DJ, Park R (2007) Cisplatin cytotoxicity of auditory cells requires secretions of proinflammatory cytokines via activation of ERK and NF-kappaB. J Assoc Res Otolaryngol 8:338–355

81. Chung WH, Boo SH, Chung MK, Lee HS, Cho YS, Hong SH (2008) Proapoptotic effects of NF-kappaB on cisplatin-induced cell death in auditory cell line. Acta Otolaryngol 128:1063–1070

82. Bortner CD, Cidlowski JA (1999) Caspase independent/dependent regulation of K(+), cell shrinkage, and mitochondrial membrane potential during lymphocyte apoptosis. J Biol Chem 274:21953–21962

83. Bortner CD, Hughes FM Jr, Cidlowski JA (1997) A primary role for K+ and Na+ efflux in the activation of apoptosis. J Biol Chem 272:32436–32442

84. Dasari S, Tchounwou PB (2014) Cisplatin in cancer therapy: molecular mechanisms of action. Eur J Pharmacol 740:364–378

85. Boonstra A, van Oudenaren A, Baert M, Leenen PJM, Savelkoul HFJ, van Steeg H, van der Horst GTJ, Hoeijmakers JHJ, Garssen J (2001) Differential ultraviolet-B-induced immunomodulation in XPA, XPC, and CSB DNA repair-deficient mice. J Investig Dermatol 117:141–146

86. Ferry KV, Hamilton TC, Johnson SW (2000) Increased nucleotide excision repair in cisplatin-resistant ovarian cancer cells: role of ERCC1-XPF. Biochem Pharmacol 60:1305–1313

87. Tornaletti S, Patrick SM, Turchi JJ, Hanawalt PC (2003) Behavior of T7 RNA polymerase and mammalian RNA polymerase II at site-specific cisplatin adducts in the template DNA. J Biol Chem 278:35791–35797

88. Guthrie OW, Li-Korotky H-S, Durrant JD, Balaban C (2008) Cisplatin induces cytoplasmic to nuclear translocation of nucleotide excision repair factors among spiral ganglion neurons. Hear Res 239:79–91

89. Caronia D, Patino-Garcia A, Milne RL, Zalacain-Diez M, Pita G, Alonso MR, Moreno LT, Sierrasesumaga-Ariznabarreta L, Benitez J, Gonzalez-Neira A (2009) Common variations in ERCC2 are associated with response to cisplatin chemotherapy and clinical outcome in osteosarcoma patients. Pharmacogenomics J 9:347–353

90. Zhang M, Liu W, Ding D, Salvi R (2003) Pifithrin-alpha suppresses p53 and protects cochlear and vestibular hair cells from cisplatin-induced apoptosis. Neuroscience 120:191–205

91. Cheng AG, Cunningham LL, Rubel EW (2005) Mechanisms of hair cell death and protection. Curr Opin Otolaryngol Head Neck Surg 13:343–348

92. Tserga E, Nandwani T, Edvall NK, Bulla J, Patel P, Canlon B, Cederroth CR, Baguley DM (2019) The genetic vulnerability to cisplatin ototoxicity: a systematic review. Sci Rep 9:3455

93. Chen HH, Kuo MT (2010) Role of glutathione in the regulation of Cisplatin resistance in cancer chemotherapy. Met Based Drugs 2010:430939

94. Peters U, Preisler-Adams S, Hebeisen A, Hahn M, Seifert E, Lanvers C, Heinecke A, Horst J, Jurgens H, Lamprecht-Dinnesen A (2000) Glutathione S-transferase genetic polymorphisms and individual sensitivity to the ototoxic effect of cisplatin. Anti-Cancer Drugs 11:639–643

95. Oldenburg J, Kraggerud SM, Cvancarova M, Lothe RA, Fossa SD (2007) Cisplatin-induced long-term hearing impairment is associated with specific glutathione s-transferase genotypes in testicular cancer survivors. J Clin Oncol 25:708–714

96. Christensen EI, Gliemann J, Moestrup SK (1992) Renal tubule gp330 is a calcium binding receptor for endocytic uptake of protein. J Histochem Cytochem 40:1481–1490

97. Hosokawa S, Hosokawa K, Ishiyama G, Ishiyama A, Lopez IA (2018) Immunohistochemical localization of megalin and cubilin in the human inner ear. Brain Res 1701:153–160

98. Tauris J, Christensen EI, Nykjaer A, Jacobsen C, Petersen CM, Ovesen T (2009) Cubilin and megalin co-localize in the neonatal inner ear. Audiol Neurootol 14:267–278

99. Konig O, Ruttiger L, Muller M, Zimmermann U, Erdmann B, Kalbacher H, Gross M, Knipper M (2008) Estrogen and the inner ear: megalin knockout mice suffer progressive hearing loss. FASEB J 22:410–417

100. Li Q, Lei F, Tang Y, Pan JS, Tong Q, Sun Y, Sheikh-Hamad D (2018) Megalin mediates plasma membrane to mitochondria cross-talk and regulates mitochondrial metabolism. Cell Mol Life Sci 75:4021–4040

101. Liu H, Pecka JL, Zhang Q, Soukup GA, Beisel KW, He DZ (2014) Characterization of transcriptomes of cochlear inner and outer hair cells. J Neurosci 34:11085–11095

102. Degl'Innocenti D, Marzocchini R, Rosati F, Cellini E, Raugei G, Ramponi G (1999) Acylphosphatase expression during macrophage differentiation and activation of U-937 cell line. Biochimie 81:1031–1035

103. Drogemoller BI, Brooks B, Critchley C, Monzon JG, Wright GEB, Liu G, Renouf DJ, Kollmannsberger CK, Bedard PL, Hayden MR, Gelmon KA, Carleton BC, Ross CJD (2018) Further investigation of the role of ACYP2 and WFS1 pharmacogenomic variants in the development of cisplatin-induced ototoxicity in testicular cancer patients. Clin Cancer Res 24:1866–1871

104. Vos HI, Guchelaar HJ, Gelderblom H, de Bont ES, Kremer LC, Naber AM, Hakobjan MH, van der Graaf WT, Coenen MJ, te Loo DM (2016) Replication of a genetic variant in ACYP2 associated with cisplatin-induced hearing loss in patients with osteosarcoma. Pharmacogenet Genomics 26:243–247

105. Xu H, Robinson GW, Huang J, Lim JY, Zhang H, Bass JK, Broniscer A, Chintagumpala M, Bartels U, Gururangan S, Hassall T, Fisher M, Cohn R, Yamashita T, Teitz T, Zuo J, Onar-Thomas A, Gajjar A, Stewart CF, Yang JJ (2015) Common variants in ACYP2 influence susceptibility to cisplatin-induced hearing loss. Nat Genet 47:263–266

106. Bhavsar AP, Gunaretnam EP, Li Y, Hasbullah JS, Carleton BC, Ross CJ (2017) Pharmacogenetic variants in TPMT alter cellular responses to cisplatin in inner ear cell lines. PLoS One 12:e0175711

107. Asadov C, Aliyeva G, Mustafayeva K (2017) Thiopurine S-methyltransferase as a pharmaco-genetic biomarker: significance of testing and review of major methods. Cardiovasc Hematol Agents Med Chem 15:23–30

108. Pussegoda K, Ross CJ, Visscher H, Yazdanpanah M, Brooks B, Rassekh SR, Zada YF, Dube MP, Carleton BC, Hayden MR (2013) Replication of TPMT and ABCC3 genetic variants highly associated with cisplatin-induced hearing loss in children. Clin Pharmacol Ther 94:243–251

109. Ross CJ, Katzov-Eckert H, Dube MP, Brooks B, Rassekh SR, Barhdadi A, Feroz-Zada Y, Visscher H, Brown AM, Rieder MJ, Rogers PC, Phillips MS, Carleton BC, Hayden MR (2009) Genetic variants in TPMT and COMT are associated with hearing loss in children receiving cisplatin chemotherapy. Nat Genet 41:1345–1349

110. Du X, Schwander M, Moresco EM, Viviani P, Haller C, Hildebrand MS, Pak K, Tarantino L, Roberts A, Richardson H, Koob G, Najmabadi H, Ryan AF, Smith RJ, Muller U, Beutler B (2008) A catechol-O-methyltransferase that is essential for auditory function in mice and humans. Proc Natl Acad Sci U S A 105:14609–14614

111. Thomas Dickey D, Muldoon LL, Kraemer DF, Neuwelt EA (2004) Protection against cisplatin-induced ototoxicity by N-acetylcysteine in a rat model. Hear Res 193:25–30

112. Choe WT, Chinosornvatana N, Chang KW (2004) Prevention of cisplatin ototoxicity using transtympanic N-acetylcysteine and lactate. Otol Neurotol 25:910–915

113. Otto WC, Brown RD, Gage-White L, Kupetz S, Anniko M, Penny JE, Henley CM (1988) Effects of cisplatin and thiosulfate upon auditory brainstem responses of guinea pigs. Hear Res 35:79–85

114. Church MW, Blakley BW, Burgio DL, Gupta AK (2004) WR-2721 (Amifostine) ameliorates cisplatin-induced hearing loss but causes neurotoxicity in hamsters: dose-dependent effects. J Assoc Res Otolaryngol 5:227–237

115. Campbell KC, Rybak LP, Meech RP, Hughes L (1996) D-methionine provides excellent protection from cisplatin ototoxicity in the rat. Hear Res 102:90–98

116. Campbell KC, Meech RP, Rybak LP, Hughes LF (2003) The effect of D-methionine on cochlear oxidative state with and without cisplatin administration: mechanisms of otoprotection. J Am Acad Audiol 14:144–156

117. Lawenda BD, Kelly KM, Ladas EJ, Sagar SM, Vickers A, Blumberg JB (2008) Should supplemental antioxidant administration be avoided during chemotherapy and radiation therapy? J Natl Cancer Inst 100:773–783

118. Viallet NR, Blakley BB, Begleiter A, Leith MK (2006) Sodium thiosulphate impairs the cytotoxic effects of cisplatin on FADU cells in culture. J Otolaryngol 35:19–21

119. Wu YJ, Muldoon LL, Neuwelt EA (2005) The chemoprotective agent N-acetylcysteine blocks cisplatin-induced apoptosis through caspase signaling pathway. J Pharmacol Exp Ther 312:424–431

120. Ekborn A, Hansson J, Ehrsson H, Eksborg S, Wallin I, Wagenius G, Laurell G (2004) High-dose Cisplatin with amifostine: ototoxicity and pharmacokinetics. Laryngoscope 114:1660–1667

121. DeVries TA, Kalkofen RL, Matassa AA, Reyland ME (2004) Protein kinase Cdelta regulates apoptosis via activation of STAT1. J Biol Chem 279:45603–45612

122. Stephanou A, Scarabelli TM, Brar BK, Nakanishi Y, Matsumura M, Knight RA, Latchman DS (2001) Induction of apoptosis and Fas receptor/Fas ligand expression by ischemia/reperfusion in cardiac myocytes requires serine 727 of the STAT-1 transcription factor but not tyrosine 701. J Biol Chem 276:28340–28347

123. Stephanou A, Scarabelli TM, Townsend PA, Bell R, Yellon D, Knight RA, Latchman DS (2002) The carboxyl-terminal activation domain of the STAT-1 transcription factor enhances ischemia/reperfusion-induced apoptosis in cardiac myocytes. FASEB J 16:1841–1843

124. Townsend PA, Scarabelli TM, Davidson SM, Knight RA, Latchman DS, Stephanou A (2004) STAT-1 interacts with p53 to enhance DNA damage-induced apoptosis. J Biol Chem 279:5811–5820

125. Stephanou A, Brar BK, Scarabelli TM, Jonassen AK, Yellon DM, Marber MS, Knight RA, Latchman DS (2000) Ischemia-induced STAT-1 expression and activation play a critical role in cardiomyocyte apoptosis. J Biol Chem 275:10002–10008

126. Schmitt NC, Rubel EW, Nathanson NM (2009) Cisplatin-induced hair cell death requires STAT1 and is attenuated by epigallocatechin gallate. J Neurosci 29:3843–3851

127. Levano S, Bodmer D (2015) Loss of STAT1 protects hair cells from ototoxicity through modulation of STAT3, c-Jun, Akt, and autophagy factors. Cell Death Dis 6:e2019

128. Stephanou A (2004) Role of STAT-1 and STAT-3 in ischaemia/reperfusion injury. J Cell Mol Med 8:519–525

129. Kaur T, Mukherjea D, Sheehan K, Jajoo S, Rybak LP, Ramkumar V (2011) Short interfering RNA against STAT1 attenuates cisplatin-induced ototoxicity in the rat by suppressing inflammation. Cell Death Dis 2:e180

130. Keithley EM, Wang X, Barkdull GC (2008) Tumor necrosis factor alpha can induce recruitment of inflammatory cells to the cochlea. Otol Neurotol 29:854–859

131. Barry SP, Townsend PA, McCormick J, Knight RA, Scarabelli TM, Latchman DS, Stephanou A (2009) STAT3 deletion sensitizes cells to oxidative stress. Biochem Biophys Res Commun 385:324–329

132. Barry SP, Townsend PA, Knight RA, Scarabelli TM, Latchman DS, Stephanou A (2010) STAT3 modulates the DNA damage response pathway. Int J Exp Pathol 91:506–514

133. Snyder M, Huang XY, Zhang JJ (2008) Identification of novel direct Stat3 target genes for control of growth and differentiation. J Biol Chem 283:3791–3798

134. Vlajkovic SM, Housley GD, Thorne PR (2009) Adenosine and the auditory system. Curr Neuropharmacol 7:246–256

135. Ford MS, Maggirwar SB, Rybak LP, Whitworth C, Ramkumar V (1997) Expression and function of adenosine receptors in the chinchilla cochlea. Hear Res 105:130–140

136. Kaur T, Borse V, Sheth S, Sheehan K, Ghosh S, Tupal S, Jajoo S, Mukherjea D, Rybak LP, Ramkumar V (2016) Adenosine A1 receptor protects against cisplatin ototoxicity by suppressing the NOX3/STAT1 inflammatory pathway in the cochlea. J Neurosci 36:3962–3977

137. Jeong HJ, Kim SJ, Moon PD, Kim NH, Kim JS, Park RK, Kim MS, Park BR, Jeong S, Um JY, Kim HM, Hong SH (2007) Antiapoptotic mechanism of cannabinoid receptor 2 agonist on cisplatin-induced apoptosis in the HEI-OC1 auditory cell line. J Neurosci Res 85:896–905

138. Martín-Saldaña S, Trinidad A, Ramil E, Sánchez-López AJ, Coronado MJ, Martínez-Martínez E, García JM, García-Berrocal JR, Ramírez-Camacho R (2016) Spontaneous cannabinoid receptor 2 (CB2) expression in the cochlea of adult albino rat and its up-regulation after cisplatin treatment. PLoS One 11:e0161954

139. Herr DR, Reolo MJY, Peh YX, Wang W, Lee C-W, Rivera R, Paterson IC, Chun J (2016) Sphingosine 1-phosphate receptor 2 (S1P2) attenuates reactive oxygen species formation and inhibits cell death: implications for otoprotective therapy. Sci Rep 6:24541

140. Wettschureck N, Offermanns S (2005) Mammalian G proteins and their cell type specific functions. Physiol Rev 85:1159–1204

141. Blair BG, Larson CA, Adams PL, Abada PB, Safaei R, Howell SB (2010) Regulation of copper transporter 2 expression by copper and cisplatin in human ovarian carcinoma cells. Mol Pharmacol 77:912–921

142. Marshak T, Steiner M, Kaminer M, Levy L, Shupak A (2014) Prevention of cisplatin-induced hearing loss by intratympanic dexamethasone: a randomized controlled study. Otolaryngol Head Neck Surg 150:983–990

143. Dias MA, Sampaio AL, Venosa AR, Meneses Ede A, Oliveira CA (2015) The chemopreventive effect of Ginkgo biloba extract 761 against cisplatin ototoxicity: a pilot study. Int Tinnitus J 19:12–19

144. Crabb SJ, Martin K, Abab J, Ratcliffe I, Thornton R, Lineton B, Ellis M, Moody R, Stanton L, Galanopoulou A, Maishman T, Geldart T, Bayne M, Davies J, Lamb C, Popat S, Joffe JK, Nutting C, Chester J, Hartley A, Thomas G, Ottensmeier C, Huddart R, King E (2017) COAST (Cisplatin ototoxicity attenuated by aspirin trial): a phase II double-blind, ran-

domised controlled trial to establish if aspirin reduces cisplatin induced hearing-loss. Eur J Cancer 87:75–83

145. Rolland V, Meyer F, Guitton MJ, Bussieres R, Philippon D, Bairati I, Leclerc M, Cote M (2019) A randomized controlled trial to test the efficacy of trans-tympanic injections of a sodium thiosulfate gel to prevent cisplatin-induced ototoxicity in patients with head and neck cancer. J Otolaryngol Head Neck Surg 48:4

146. Freyer DR, Chen L, Krailo MD, Knight K, Villaluna D, Bliss B, Pollock BH, Ramdas J, Lange B, Van Hoff D, VanSoelen ML, Wiernikowski J, Neuwelt EA, Sung L (2017) Effects of sodium thiosulfate versus observation on development of cisplatin-induced hearing loss in children with cancer (ACCL0431): a multicentre, randomised, controlled, open-label, phase 3 trial. Lancet Oncol 18:63–74

147. Zuur CL, Simis YJ, Lansdaal PE, Hart AA, Schornagel JH, Dreschler WA, Rasch CR, Balm AJ (2007) Ototoxicity in a randomized phase III trial of intra-arterial compared with intravenous cisplatin chemoradiation in patients with locally advanced head and neck cancer. J Clin Oncol 25:3759–3765

148. Ishikawa E, Sugimoto H, Hatano M, Nakanishi Y, Tsuji A, Endo K, Kondo S, Wakisaka N, Murono S, Ito M, Yoshizaki T (2015) Protective effects of sodium thiosulfate for cisplatin-mediated ototoxicity in patients with head and neck cancer. Acta Otolaryngol 135:919–924

149. Madasu R, Ruckenstein MJ, Leake F, Steere E, Robbins KT (1997) Ototoxic effects of supradose cisplatin with sodium thiosulfate neutralization in patients with head and neck cancer. Arch Otolaryngol Head Neck Surg 123:978–981

150. Womack AM, Hayes-Jordan A, Pratihar R, Barringer DA, Hall JH Jr, Gidley PW, Lewin JS (2014) Evaluation of ototoxicity in patients treated with hyperthermic intraperitoneal chemotherapy (HIPEC) with cisplatin and sodium thiosulfate. Ear Hear 35:e243–e247

151. van Rijswijk RE, Hoekman K, Burger CW, Verheijen RH, Vermorken JB (1997) Experience with intraperitoneal cisplatin and etoposide and i.v. sodium thiosulphate protection in ovarian cancer patients with either pathologically complete response or minimal residual disease. Ann Oncol 8:1235–1241

152. Sarafraz Z, Ahmadi A, Daneshi A (2018) Transtympanic injections of N-acetylcysteine and dexamethasone for prevention of cisplatin-induced ototoxicity: double blind randomized clinical trial. Int Tinnitus J 22:40–45

153. Yoo J, Hamilton SJ, Angel D, Fung K, Franklin J, Parnes LS, Lewis D, Venkatesan V, Winquist E (2014) Cisplatin otoprotection using transtympanic L-N-acetylcysteine: a pilot randomized study in head and neck cancer patients. Laryngoscope 124:E87–E94

154. Gurney JG, Bass JK, Onar-Thomas A, Huang J, Chintagumpala M, Bouffet E, Hassall T, Gururangan S, Heath JA, Kellie S, Cohn R, Fisher MJ, Panandiker AP, Merchant TE, Srinivasan A, Wetmore C, Qaddoumi I, Stewart CF, Armstrong GT, Broniscer A, Gajjar A (2014) Evaluation of amifostine for protection against cisplatin-induced serious hearing loss in children treated for average-risk or high-risk medulloblastoma. Neuro-Oncology 16:848–855

155. Kemp G, Rose P, Lurain J, Berman M, Manetta A, Roullet B, Homesley H, Belpomme D, Glick J (1996) Amifostine pretreatment for protection against cyclophosphamide-induced and cisplatin-induced toxicities: results of a randomized control trial in patients with advanced ovarian cancer. J Clin Oncol 14:2101–2112

156. Planting AS, Catimel G, de Mulder PH, de Graeff A, Hoppener F, Verweij J, Oster W, Vermorken JB (1999) Randomized study of a short course of weekly cisplatin with or without amifostine in advanced head and neck cancer. EORTC Head and Neck Cooperative Group. Ann Oncol 10:693–700

157. Fisher MJ, Lange BJ, Needle MN, Janss AJ, Shu HK, Adamson PC, Phillips PC (2004) Amifostine for children with medulloblastoma treated with cisplatin-based chemotherapy. Pediatr Blood Cancer 43:780–784

158. Marina N, Chang KW, Malogolowkin M, London WB, Frazier AL, Womer RB, Rescorla F, Billmire DF, Davis MM, Perlman EJ, Giller R, Lauer SJ, Olson TA, Group Children's

Oncology (2005) Amifostine does not protect against the ototoxicity of high-dose cis-platin combined with etoposide and bleomycin in pediatric germ-cell tumors: a Children's Oncology Group Study. Cancer 104:841–847

159. Katzenstein HM, Chang KW, Krailo M, Chen Z, Finegold MJ, Rowland J, Reynolds M, Pappo A, London WB, Malogolowkin M, Group Children's Oncology (2009) Amifostine does not prevent platinum-induced hearing loss associated with the treatment of children with hepatoblastoma: a report of the Intergroup Hepatoblastoma Study P9645 as a part of the Children's Oncology Group. Cancer 115:5828–5835

160. Prinja S, Singh G, Vashisth M, Arora T (2016) Protective role of calcium channel blocker flunarizine on cisplatin induced ototoxicity: a clinical study. Int J Contemp Med Res 3:1290–1292

161. Villani V, Zucchella C, Cristalli G, Galie E, Bianco F, Giannarelli D, Carpano S, Spriano G, Pace A (2016) Vitamin E neuroprotection against cisplatin ototoxicity: preliminary results from a randomized, placebo-controlled trial. Head Neck 38(Suppl 1):E2118–E2121

162. Weijl NI, Elsendoorn TJ, Lentjes EG, Hopman GD, Wipkink-Bakker A, Zwinderman AH, Cleton FJ, Osanto S (2004) Supplementation with antioxidant micronutrients and chemotherapy-induced toxicity in cancer patients treated with cisplatin-based chemotherapy: a randomised, double-blind, placebo-controlled study. Eur J Cancer 40:1713–1723

163. de Graaf TW, de Jong S, de Vries EG, Mulder NH (1997) Expression of proteins correlated with the unique cisplatin-sensitivity of testicular cancer. Anticancer Res 17:369–375

164. Kikuchi Y, Miyauchi M, Kizawa I, Oomori K, Kato K (1986) Establishment of a cisplatin-resistant human ovarian cancer cell line. J Natl Cancer Inst 77:1181–1185

165. von Eyben FE, Blaabjerg O, Madsen EL, Petersen PH, Smith-Sivertsen C, Gullberg B (1992) Serum lactate dehydrogenase isoenzyme 1 and tumour volume are indicators of response to treatment and predictors of prognosis in metastatic testicular germ cell tumours. Eur J Cancer 28:410–415

166. Spector GJ, Carr C (1974) The electron transport system in the cochlear hair cell: the ultra-structural cytochemistry of respiratory enzymes in hair cell mitochondria of the guinea pig. Laryngoscope 84:1673–1706

167. Haupt H, Scheibe F, Bergmann K (1983) [Total lactate dehydrogenase activity of peri-lymph, plasma and cerebrospinal fluid in unstressed and noise stressed guinea pigs]. Arch Otorhinolaryngol 238:77–85

168. Scheibe F, Haupt H, Rothe E, Hache U (1981) [Lactate and pyruvate concentrations in peri-lymph, blood, and cerebrospinal fluid of guinea pigs]. Arch Otorhinolaryngol 232:81–89

169. Hannemann J, Baumann K (1988) Inhibition of lactate-dehydrogenase by cisplatin and other platinum-compounds: enzyme leakage of LDH is not a suitable method to measure platinum-compound-induced kidney cell damage in vitro. Res Commun Chem Pathol Pharmacol 60:371–379

170. Zhang JG, Lindup WE (1996) Differential effects of cisplatin on the production of NADH-dependent superoxide and the activity of antioxidant enzymes in rat renal cortical slices in vitro. Pharmacol Toxicol 79:191–198

171. Crini G (2014) Review: a history of cyclodextrins. Chem Rev 114:10940–10975

172. Loftsson T, Hreinsdottir D, Masson M (2005) Evaluation of cyclodextrin solubilization of drugs. Int J Pharm 302:18–28

173. Conceicao J, Adeoye O, Cabral-Marques HM, Lobo JMS (2018) Cyclodextrins as drug carriers in pharmaceutical technology: the state of the art. Curr Pharm Des 24:1405–1433

174. Davis ME, Brewster ME (2004) Cyclodextrin-based pharmaceutics: past, present and future. Nat Rev Drug Discov 3:1023–1035

175. Loftsson T, Brewster ME (2010) Pharmaceutical applications of cyclodextrins: basic science and product development. J Pharm Pharmacol 62:1607–1621

176. Zhang J, Ma PX (2013) Cyclodextrin-based supramolecular systems for drug delivery: recent progress and future perspective. Adv Drug Deliv Rev 65:1215–1233

177. Ohtani Y, Irie T, Uekama K, Fukunaga K, Pitha J (1989) Differential effects of alpha-, beta- and gamma-cyclodextrins on human erythrocytes. Eur J Biochem 186:17–22
178. Zidovetzki R, Levitan I (2007) Use of cyclodextrins to manipulate plasma membrane cholesterol content: evidence, misconceptions and control strategies. Biochim Biophys Acta 1768:1311–1324
179. Gould S, Scott RC (2005) 2-Hydroxypropyl-beta-cyclodextrin (HP-beta-CD): a toxicology review. Food Chem Toxicol 43:1451–1459
180. Irie T, Uekama K (1997) Pharmaceutical applications of cyclodextrins. III. Toxicological issues and safety evaluation. J Pharm Sci 86:147–162
181. Otero-Espinar FJ, Luzardo-Alvarez A, Blanco-Mendez J (2010) Cyclodextrins: more than pharmaceutical excipients. Mini Rev Med Chem 10:715–725
182. Hastings, C (2010) Request for intrathecal delivery of HPBCD for niemann pick Type C patients, Caroline Hastings, M.D. Principal Investigator Department of Pediatric Hematology Oncology Children's Hospital & Research Center Oakland Submission Date to FDA.
183. Megías-Vericat JE, Company-Albir MJ, García-Robles AA, Poveda JL (2017) Use of 2-hydroxypropyl-beta-cyclodextrin for Niemann-Pick type C disease. In: Dhingra N, Arora P (eds) Cyclodextrin—a versatile ingredient. IntechOpen, London
184. Crumling MA, Liu L, Thomas PV, Benson J, Kanicki A, Kabara L, Halsey K, Dolan D, Duncan RK (2012) Hearing loss and hair cell death in mice given the cholesterol-chelating agent hydroxypropyl-beta-cyclodextrin. PLoS One 7:e53280
185. Ory DS, Ottinger EA, Farhat NY, King KA, Jiang X, Weissfeld L, Berry-Kravis E, Davidson CD, Bianconi S, Keener LA, Rao R, Soldatos A, Sidhu R, Walters KA, Xu X, Thurm A, Solomon B, Pavan WJ, Machielse BN, Kao M, Silber SA, McKew JC, Brewer CC, Vite CH, Walkley SU, Austin CP, Porter FD (2017) Intrathecal 2-hydroxypropyl-beta-cyclodextrin decreases neurological disease progression in Niemann-Pick disease, type C1: a non-randomised, open-label, phase 1-2 trial. Lancet 390:1758–1768
186. Ward S, O'Donnell P, Fernandez S, Vite CH (2010) 2-Hydroxypropyl-beta-cyclodextrin raises hearing threshold in normal cats and in cats with Niemann-Pick type C disease. Pediatr Res 68:52–56
187. Cronin S, Lin A, Thompson K, Hoenerhoff M, Duncan RK (2015) Hearing loss and otopathology following systemic and intracerebroventricular delivery of 2-hydroxypropyl-beta-cyclodextrin. J Assoc Res Otolaryngol 16:599–611
188. Nakashima T, Sone M, Teranishi M, Yoshida T, Terasaki H, Kondo M, Yasuma T, Wakabayashi T, Nagatani T, Naganawa S (2012) A perspective from magnetic resonance imaging findings of the inner ear: relationships among cerebrospinal, ocular and inner ear fluids. Auris Nasus Larynx 39:345–355
189. Nguyen T-VN, Brownell WE (1998) Contribution of membrane cholesterol to outer hair cell lateral wall stiffness. Otolaryngol Head Neck Surg 119:14–20
190. Thomas PV, Cheng AL, Colby CC, Liu L, Patel CK, Josephs L, Duncan RK (2014) Localization and proteomic characterization of cholesterol-rich membrane microdomains in the inner ear. J Proteome 103:178–193
191. Kamar RI, Organ-Darling LE, Raphael RM (2012) Membrane cholesterol strongly influences confined diffusion of prestin. Biophys J 103:1627–1636
192. Rajagopalan L, Organ-Darling LE, Liu H, Davidson AL, Raphael RM, Brownell WE, Pereira FA (2010) Glycosylation regulates prestin cellular activity. J Assoc Res Otolaryngol 11:39–51
193. Mount DB, Romero MF (2004) The SLC26 gene family of multifunctional anion exchangers. Pflugers Arch 447:710–721
194. Muallem D, Ashmore J (2006) An anion antiporter model of prestin, the outer hair cell motor protein. Biophys J 90:4035–4045
195. Cheatham MA, Edge RM, Homma K, Leserman EL, Dallos P, Zheng J (2015) Prestin-dependence of outer hair cell survival and partial rescue of outer hair cell loss in Prestin V499G/Y501H knockin mice. PLoS One 10:e0145428

196. Dallos P (2008) Cochlear amplification, outer hair cells and prestin. Curr Opin Neurobiol 18:370–376
197. Liberman MC, Gao J, He DZ, Wu X, Jia S, Zuo J (2002) Prestin is required for electromotility of the outer hair cell and for the cochlear amplifier. Nature 419:300–304
198. Wu X, Gao J, Guo Y, Zuo J (2004) Hearing threshold elevation precedes hair-cell loss in prestin knockout mice. Brain Res Mol Brain Res 126:30–37
199. Takahashi S, Homma K, Zhou Y, Nishimura S, Duan C, Chen J, Ahmad A, Cheatham MA, Zheng J (2016) Susceptibility of outer hair cells to cholesterol chelator 2-hydroxypropyl-beta-cyclodextrine is prestin-dependent. Sci Rep 6:21973
200. Canis M, Schmid J, Olzowy B, Jahn K, Strupp M, Berghaus A, Suckfuell M (2009) The influence of cholesterol on the motility of cochlear outer hair cells and the motor protein prestin. Acta Otolaryngol 129:929–934
201. Rajagopalan L, Greeson JN, Xia A, Liu H, Sturm A, Raphael RM, Davidson AL, Oghalai JS, Pereira FA, Brownell WE (2007) Tuning of the outer hair cell motor by membrane cholesterol. J Biol Chem 282:36659–36670
202. Brownell WE, Jacob S, Hakizimana P, Ulfendahl M, Fridberger A (2011) Decreasing outer hair cell membrane cholesterol increases cochlear electromechanics. AIP Conf Proc 1403:148–153
203. Sfondouris J, Rajagopalan L, Pereira FA, Brownell WE (2008) Membrane composition modulates prestin-associated charge movement. J Biol Chem 283:22473–22481
204. Zhou Y, Takahashi S, Homma K, Duan C, Zheng J, Cheatham MA, Zheng J (2018) The susceptibility of cochlear outer hair cells to cyclodextrin is not related to their electromotile activity. Acta Neuropathol Commun 6:98
205. Lee MY, Kabara LL, Swiderski DL, Raphael Y, Duncan RK, Kim YH (2019) ROS scavenger, ebselen, has no preventive effect in new hearing loss model using a cholesterol-chelating agent. J Audiol Otol 23:69–75
206. Ziolkowski W, Szkatula M, Nurczyk A, Wakabayashi T, Kaczor JJ, Olek RA, Knap N, Antosiewicz J, Wieckowski MR, Wozniak M (2010) Methyl-beta-cyclodextrin induces mitochondrial cholesterol depletion and alters the mitochondrial structure and bioenergetics. FEBS Lett 584:4606–4610
207. Lichtenhan JT, Hirose K, Buchman CA, Duncan RK, Salt AN (2017) Direct administration of 2-hydroxypropyl-beta-cyclodextrin into guinea pig cochleae: effects on physiological and histological measurements. PLoS One 12:e0175236
208. Purcell EK, Liu L, Thomas PV, Duncan RK (2011) Cholesterol influences voltage-gated calcium channels and BK-type potassium channels in auditory hair cells. PLoS One 6:e26289
209. Kumana CR, Yuen KY (1994) Parenteral aminoglycoside therapy. Selection, administration and monitoring. Drugs 47:902–913
210. Fourmy D, Recht MI, Blanchard SC, Puglisi JD (1996) Structure of the A site of Escherichia coli 16S ribosomal RNA complexed with an aminoglycoside antibiotic. Science 274:1367–1371
211. Mingeot-Leclercq MP, Glupczynski Y, Tulkens PM (1999) Aminoglycosides: activity and resistance. Antimicrob Agents Chemother 43:727–737
212. Hinshaw HC, Feldman WH, Pfuetze KH (1946) Treatment of tuberculosis with streptomycin; a summary of observations on one hundred cases. J Am Med Assoc 132:778–782
213. Selimoglu E (2007) Aminoglycoside-induced ototoxicity. Curr Pharm Des 13:119–126
214. Fausti SA, Henry JA, Schaffer HI, Olson DJ, Frey RH, McDonald WJ (1992) High-frequency audiometric monitoring for early detection of aminoglycoside ototoxicity. J Infect Dis 165:1026–1032
215. Kusunoki T, Cureoglu S, Schachern PA, Sampaio A, Fukushima H, Oktay MF, Paparella MM (2004) Effects of aminoglycoside administration on cochlear elements in human temporal bones. Auris Nasus Larynx 31:383–388
216. Petersen L, Rogers C (2015) Aminoglycoside-induced hearing deficits—a review of cochlear ototoxicity. S Afr Fam Pract 57:77–82

217. Lerner SA, Matz GJ (1979) Suggestions for monitoring patients during treatment with aminoglycoside antibiotics. Otolaryngol Head Neck Surg 87:222–228

218. Moore RD, Smith CR, Lietman PS (1984) Risk factors for the development of auditory toxicity in patients receiving aminoglycosides. J Infect Dis 149:23–30

219. Fee WE Jr (1980) Aminoglycoside ototoxicity in the human. Laryngoscope 90:1–19

220. Fausti SA, Frey RH, Henry JA, Olson DJ, Schaffer HI (1993) High-frequency testing techniques and instrumentation for early detection of ototoxicity. J Rehabil Res Dev 30:333–341

221. Hotz MA, Harris FP, Probst R (1994) Otoacoustic emissions: an approach for monitoring aminoglycoside-induced ototoxicity. Laryngoscope 104:1130–1134

222. de Jager P, van Altena R (2002) Hearing loss and nephrotoxicity in long-term aminoglycoside treatment in patients with tuberculosis. Int J Tuberc Lung Dis 6:622–627

223. Duggal P, Sarkar M (2007) Audiologic monitoring of multi-drug resistant tuberculosis patients on aminoglycoside treatment with long term follow-up. BMC Ear Nose Throat Disord 7:5

224. Sturdy A, Goodman A, Jose RJ, Loyse A, O'Donoghue M, Kon OM, Dedicoat MJ, Harrison TS, John L, Lipman M, Cooke GS (2011) Multidrug-resistant tuberculosis (MDR-TB) treatment in the UK: a study of injectable use and toxicity in practice. J Antimicrob Chemother 66:1815–1820

225. Al-Malky G, Suri R, Dawson SJ, Sirimanna T, Kemp D (2011) Aminoglycoside antibiotics cochleotoxicity in paediatric cystic fibrosis (CF) patients: a study using extended high-frequency audiometry and distortion product otoacoustic emissions. Int J Audiol 50:112–122

226. Ramma L, Ibekwe TS (2012) Cochleo-vestibular clinical findings among drug resistant tuberculosis patients on therapy—a pilot study. Int Arch Med 5:3

227. Harris T, Bardien S, Schaaf HS, Petersen L, De Jong G, Fagan JJ (2012) Aminoglycoside-induced hearing loss in HIV-positive and HIV-negative multidrug-resistant tuberculosis patients. S Afr Med J 102:363–366

228. Rybak LP, Schacht J (2008) Drug-induced hearing loss. Springer, New York

229. Meyerhoff WL, Maale GE, Yellin W, Roland PS (1989) Audiologic threshold monitoring of patients receiving ototoxic drugs. Preliminary report. Ann Otol Rhinol Laryngol 98:950–954

230. Bath AP, Walsh RM, Bance ML, Rutka JA (1999) Ototoxicity of topical gentamicin preparations. Laryngoscope 109:1088–1093

231. Saleh P, Abbasalizadeh S, Rezaeian S, Naghavi-Behzad M, Piri R, Pourfeizi HH (2016) Gentamicin-mediated ototoxicity and nephrotoxicity: a clinical trial study. Niger Med J 57:347–352

232. Stankowicz MS, Ibrahim J, Brown DL (2015) Once-daily aminoglycoside dosing: an update on current literature. Am J Health Syst Pharm 72:1357–1364

233. Hamasaki K, Rando RR (1997) Specific binding of aminoglycosides to a human rRNA construct based on a DNA polymorphism which causes aminoglycoside-induced deafness. Biochemistry 36:12323–12328

234. O'Sullivan ME, Perez A, Lin R, Sajjadi A, Ricci AJ, Cheng AG (2017) Towards the prevention of aminoglycoside-related hearing loss. Front Cell Neurosci 11:325

235. Usami S, Abe S, Shinkawa H, Kimberling WJ (1998) Sensorineural hearing loss caused by mitochondrial DNA mutations: special reference to the A1555G mutation. J Commun Disord 31:423–434; quiz 34–5

236. Hailey DW, Esterberg R, Linbo TH, Rubel EW, Raible DW (2017) Fluorescent aminoglycosides reveal intracellular trafficking routes in mechanosensory hair cells. J Clin Invest 127:472–486

237. Hashino E, Shero M, Salvi RJ (1997) Lysosomal targeting and accumulation of aminoglycoside antibiotics in sensory hair cells. Brain Res 777:75–85

238. Jiang M, Karasawa T, Steyger PS (2017) Aminoglycoside-induced cochleotoxicity: a review. Front Cell Neurosci 11:308

239. Dai CF, Mangiardi D, Cotanche DA, Steyger PS (2006) Uptake of fluorescent gentamicin by vertebrate sensory cells in vivo. Hear Res 213:64–78

240. Zheng J, Dai C, Steyger PS, Kim Y, Vass Z, Ren T, Nuttall AL (2003) Vanilloid receptors in hearing: altered cochlear sensitivity by vanilloids and expression of TRPV1 in the organ of corti. J Neurophysiol 90:444–455
241. Kros CJ, Marcotti W, van Netten SM, Self TJ, Libby RT, Brown SD, Richardson GP, Steel KP (2002) Reduced climbing and increased slipping adaptation in cochlear hair cells of mice with Myo7a mutations. Nat Neurosci 5:41–47
242. Vu AA, Nadaraja GS, Huth ME, Luk L, Kim J, Chai R, Ricci AJ, Cheng AG (2013) Integrity and regeneration of mechanotransduction machinery regulate aminoglycoside entry and sensory cell death. PLoS One 8:e54794
243. Lanvers-Kaminsky C, Ciarimboli G (2017) Pharmacogenetics of drug-induced ototoxicity caused by aminoglycosides and cisplatin. Pharmacogenomics 18:1683–1695
244. Nguyen T, Jeyakumar A (2019) Genetic susceptibility to aminoglycoside ototoxicity. Int J Pediatr Otorhinolaryngol 120:15–19
245. Zhao H, Li R, Wang Q, Yan Q, Deng JH, Han D, Bai Y, Young WY, Guan MX (2004) Maternally inherited aminoglycoside-induced and nonsyndromic deafness is associated with the novel C1494T mutation in the mitochondrial 12S rRNA gene in a large Chinese family. Am J Hum Genet 74:139–152
246. Gao Z, Chen Y, Guan MX (2017) Mitochondrial DNA mutations associated with aminoglycoside induced ototoxicity. J Otol 12:1–8
247. Huang S, Xiang G, Kang D, Wang C, Kong Y, Zhang X, Liang S, Mitchelson K, Xing W, Dai P (2015) Rapid identification of aminoglycoside-induced deafness gene mutations using multiplex real-time polymerase chain reaction. Int J Pediatr Otorhinolaryngol 79:1067–1072
248. American Academy of Audiology (2009) Position statement and clinical practice guidelines. Ototoxicity monitoring, Audiology (2014). Available at: https://www.audiology.org/publications-resources/document-library/ototoxicitymonitoring. (Accessed: 21st April 2019)
249. Li H, Steyger PS (2009) Synergistic ototoxicity due to noise exposure and aminoglycoside antibiotics. Noise Health 11:26–32
250. Campbell KCM, Le Prell CG (2018) Drug-induced ototoxicity: diagnosis and monitoring. Drug Saf 41:451–464
251. Campbell KC, Meech RP, Klemens JJ, Gerberi MT, Dyrstad SS, Larsen DL, Mitchell DL, El-Azizi M, Verhulst SJ, Hughes LF (2007) Prevention of noise- and drug-induced hearing loss with D-methionine. Hear Res 226:92–103
252. Kranzer K, Elamin WF, Cox H, Seddon JA, Ford N, Drobniewski F (2015) A systematic review and meta-analysis of the efficacy and safety of N-acetylcysteine in preventing aminoglycoside-induced ototoxicity: implications for the treatment of multidrug-resistant TB. Thorax 70:1070–1077
253. Fetoni AR, Sergi B, Ferraresi A, Paludetti G, Troiani D (2004) alpha-Tocopherol protective effects on gentamicin ototoxicity: an experimental study. Int J Audiol 43:166–171
254. Conlon BJ, Aran JM, Erre JP, Smith DW (1999) Attenuation of aminoglycoside-induced cochlear damage with the metabolic antioxidant alpha-lipoic acid. Hear Res 128:40–44
255. Fetoni AR, Eramo SL, Rolesi R, Troiani D, Paludetti G (2012) Antioxidant treatment with coenzyme Q-ter in prevention of gentamycin ototoxicity in an animal model. Acta Otorhinolaryngol Ital 32:103–110
256. Jung HW, Chang SO, Kim CS, Rhee CS, Lim DH (1998) Effects of Ginkgo biloba extract on the cochlear damage induced by local gentamicin installation in guinea pigs. J Korean Med Sci 13:525–528
257. Feldman L, Efrati S, Eviatar E, Abramsohn R, Yarovoy I, Gersch E, Averbukh Z, Weissgarten J (2007) Gentamicin-induced ototoxicity in hemodialysis patients is ameliorated by N-acetylcysteine. Kidney Int 72:359–363
258. Sha SH, Qiu JH, Schacht J (2006) Aspirin to prevent gentamicin-induced hearing loss. N Engl J Med 354:1856–1857
259. Hammill TL, Campbell KC (2018) Protection for medication-induced hearing loss: the state of the science. Int J Audiol 57:S67–S75

260. Huth ME, Ricci AJ, Cheng AG. Mechanisms of aminoglycoside ototoxicity and targets of hair cell protection. Int J Otolaryngol.2011;2011:937861

261. Alharazneh A, Luk L, Huth M, Monfared A, Steyger PS, Cheng AG, Ricci AJ (2011) Functional hair cell mechanotransducer channels are required for aminoglycoside ototoxicity. PLoS One 6:e22347

262. Dai P, Liu X, Han D, Qian Y, Huang D, Yuan H, Li W, Yu F, Zhang R, Lin H, He Y, Yu Y, Sun Q, Qin H, Li R, Zhang X, Kang D, Cao J, Young WY, Guan MX (2006) Extremely low penetrance of deafness associated with the mitochondrial 12S rRNA mutation in 16 Chinese families: implication for early detection and prevention of deafness. Biochem Biophys Res Commun 340:194–199

263. Imamura S, Adams JC (2003) Distribution of gentamicin in the guinea pig inner ear after local or systemic application. J Assoc Res Otolaryngol 4:176–195

264. Kitahara T, Li HS, Balaban CD (2005) Changes in transient receptor potential cation channel superfamily V (TRPV) mRNA expression in the mouse inner ear ganglia after kanamycin challenge. Hear Res 201:132–144

265. Matt T, Ng CL, Lang K, Sha SH, Akbergenov R, Shcherbakov D, Meyer M, Duscha S, Xie J, Dubbaka SR, Perez-Fernandez D, Vasella A, Ramakrishnan V, Schacht J, Bottger EC (2012) Dissociation of antibacterial activity and aminoglycoside ototoxicity in the 4-monosubstituted 2-deoxystreptamine apramycin. Proc Natl Acad Sci U S A 109:10984–10989

266. Huth ME, Han KH, Sotoudeh K, Hsieh YJ, Effertz T, Vu AA, Verhoeven S, Hsieh MH, Greenhouse R, Cheng AG, Ricci AJ (2015) Designer aminoglycosides prevent cochlear hair cell loss and hearing loss. J Clin Invest 125:583–592

Approaches to Regenerate Hair Cell and Spiral Ganglion Neuron in the Inner Ear

Muhammad Waqas and Renjie Chai

Abbreviations

AAV	Adeno-associated virus
Ad	Adenovirus
BDNF	Brain-derived neurotrophic factor
GDNF	Glial-derived neurotrophic factor
HCs	Hair cells
iPSCs	Inducible pluripotent stem cells
NT3	Neurotrophin 3
SCs	Supporting cells
SGNs	Spiral ganglion neurons
SNHL	Sensorineural hearing loss

1   Introduction

The inner ear is one of the most elegant anatomical structures in the body and the sensory cells within are extremely sensitive to mechanical stimulation. It contains three primary subdivisions optimized for different functions: the cochlea for sound (acoustic) perception, the vestibular otolithic organs (saccule and utricle) to detect linear acceleration, and three semicircular canals to detect head rotations in three different planes.

M. Waqas
Department of Biotechnology, Federal Urdu University of Arts, Science and Technology, Gulshan-e-Iqbal Campus, Karachi, Pakistan

MOE Key Laboratory of Developmental Genes and Human Disease, Institute of Life Sciences, Southeast University, Nanjing, China

R. Chai (✉)
MOE Key Laboratory of Developmental Genes and Human Disease, Institute of Life Sciences, Southeast University, Nanjing, China

Co-Innovation Center of Neuroregeneration, Nantong University, Nantong, China
e-mail: renjiec@seu.edu.cn

© Springer Nature Switzerland AG 2020
S. Pucheu et al. (eds.), *New Therapies to Prevent or Cure Auditory Disorders*,
https://doi.org/10.1007/978-3-030-40413-0_4

The sensory hair cells (HC) in the cochlea are responsible for transducing incoming sound vibrations, such as speech and music, into electrical signals. The inner hair cells (IHCs) are responsible for transmitting these electrical signals to the spiral ganglion neurons (SGN) located in the modiolus of the cochlea. The axons of the SGNs collectively give rise to the auditory nerve. The all-or-none spike discharges from the auditory nerve fibers are then relayed to the cochlear nucleus in the brainstem. The cochlear sensory epithelium comprises one row of IHCs and three rows of outer hair cells (OHCs), which spiral around the modiolus, in a cork-screw pattern, from the basal, high-frequency region to the low-frequency area in the apex. The IHC and OHC located in the organ of Corti on the basilar membrane are surrounded by the multiple layers of supporting cells (SCs) (Fig. 1).

There is a division of labor amongst the two types of sensory hair cells. The OHCs, which are electromotile and act like a motor, are responsible for amplifying the incoming sounds by enhancing the vibrations of the basilar membrane. The IHCs, which behave like a true sensory cell, are mainly responsible for converting the mechanical vibrations of the basilar membrane into a graded electrical neural signal that stimulates the release of neurotransmitter, putatively glutamate, from the basal pole of the IHC onto the peripheral afferent terminal of the SGN. The spike discharges generated in the SGNs are relayed through the auditory nerve to the central auditory system [1–5].

Due to their exquisite sensitivity to acoustic stimulation and delicate structure, the HCs are easily damaged by intense sounds and certain types of ototoxic damage, such as aminoglycoside antibiotics and ototoxic platinum-based anticancer drugs. As compared to the IHCs, OHCs are more prone to damage. The most common and critical histopathological findings in hearing loss patients are the loss of HCs and the reduced number of SGNs [6, 7]. The most common forms of HC death result from exposure to loud noise, aminoglycosides antibiotics, chemotherapy drugs, aging, viral infections, and inherited disorders [8–10]. Intense noise exposures and ototoxic drugs also provoke the loss of specialized synaptic structures between IHCs and SGNs, which can lead to subsequent degeneration of SGN [11, 12].

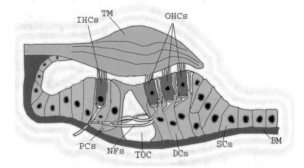

Fig. 1 Schematic cross section of the mammalian organ of Corti that shows the sensory epithelium which resides on the basilar membrane. *TM* tectorial membrane, *IHCs* inner hair cells, *OHCs* outer hair cells, *BM* basilar membrane, *SCs* supporting cells, *DCs* deiters cells, *TOC* tunnel of Corti, *NFs* nerve fibres, *PCs* pillar cells

1.1 Global Estimates and Impact of Hearing Impairment

Recent estimates of hearing loss burden indicate that around 466 million people worldwide experience some form of hearing loss. Surprisingly, among those with hearing impairment, 34 million are children. In addition, there has been a steady rise in the hearing impaired population and the prevalence of hearing impairment is expected to reach 900 million by the end of 2050 [13]. Hearing loss directly affects the person's ability to communicate with others making it extremely difficult to deal with day-to-day life situations. Severe hearing loss can have disastrous social and mental health consequences as these individuals become isolated in the world of communicators [14–17]. Hearing loss can have a significant impact on the social, emotional, and economic conditions of an individual, particularly children who are more at risk for experiencing a delay in language development [18]. More recent studies have found that hearing loss is a risk factor that has a significant association with cognitive disorder and dementia in adults and these individuals are more susceptible to accidental trauma [19–21].

1.2 Categories of Hearing Loss

There are three types of hearing impairment, conductive, sensorineural, and mixed hearing loss. Conductive hearing loss results from damage or disruption of the outer or middle ear such as the middle ear ossicles. In contrast, sensorineural hearing loss (SNHL), the most common type of permanent deafness, typically results from damage to the HC and SGN. The root cause of SNHL is always lying in the inner ear and it can be mild, moderate, severe, or profound; the later forms are difficult to treat with a hearing aid. Mixed hearing loss is caused by a combination of conductive damage and SNHL.

While most forms of conductive hearing loss may completely recover or may be surgically treated, SNHL is permanent due to the inability of HCs and SGNs to repair themselves or regenerate. While mild and moderate SNHLs can be treated with hearing aids, treatment of profound hearing loss involves cochlear implants, which bypass the missing HCs, and evoked auditory sensations by delivering electrical stimulation to the surviving SGNs located at different tonotoptic regions of the cochlea [22]. Cochlear implants have proved most beneficial to those that develop hearing loss postlingually; they are not recommended for those with prelingual deafness. Cochlear implants can provide significant benefits to postlingually deafened children under the age of 6, but little benefit to prelinquinally deafened children. Cochlear implants give the patient access to sound and aids them to acquire spoken language [23, 24]. However, the quality of the sounds perceived through the implant is inferior to that from a normal hearing ear. Deficits in sound perception through a cochlear implant are most notable when listening to music or to speech in noisy environments [25, 26]. To overcome these deficiencies, there has been a growing interest in trying to fully restore hearing using different biological approaches with the goal of restoring/regenerating the HC, supporting cells, and SGNs in individuals with severe to profound SNHL.

1.3   Native HCs Regeneration Capacity in Mammals and Non-mammals

Broadly speaking, regeneration refers to mechanisms of renewal, restoration, repair, and development that allow cells to reestablish normal or nearly normal function following damage. The story of HCs regeneration in the inner ear began approximately 30 years ago with multiple classical studies demonstrating that the residual supporting cells (SCs) of the avian basilar papilla and utricle serve as a reservoir to regenerate HCs and rebuild the sensory epithelium and restore essentially normal hearing after damage [27–30]. Extensive investigations showed that SCs convert into HCs by two different mechanisms. One is through mitotic regeneration, where SCs first divide and then one member of the mitotic pair differentiates to form an HC, while the other becomes an SC. The other more prevalent route is by transdifferentiation, a process by which the SCs directly differentiate into HCs without any prior cell division [31].

   In contrast, the mature mammalian cochlea is unable to restore hearing function due to the lack of both mitotic and non-mitotic HCs regeneration capacity [6, 32]. However, the mature mammalian vestibular system, responsible for maintaining body equilibrium, has a limited regenerative ability to produce new HCs by transdifferentiation of SCs [33–37]. Similarly, the neonatal mammalian cochlea retains limited HC regeneration potential and harbors a population of SCs that differentiate to form new HCs in vitro [38–41]. When these SCs are isolated from *p27Kip* transgenic mice using flow cytometry and then cultured in vitro, they are able to proliferate and differentiate into HCs [42]. These resident SCs from the neonatal cochlea can also form spheres and clonal colonies from a single cell and subsequently differentiate to form new HCs in vitro culture [41]. Thus, the SCs found in the neonatal and mature utricle display pluripotent stem cell characteristic [41, 43].

   Recent studies revealed that the specific subpopulation of pluripotent cochlear SCs express *Lgr5*, a stem cell marker [44]. *Lgr5+* SCs represent an enriched population of progenitors in the inner ear; they have a significant capacity to proliferate and differentiate into HCs via mitotic and/or direct transdifferentiation as compared to *Lgr5−* SCs [45–48]. To understand the role of *Lgr5+* SCs in vivo, *Pou4f3DTR* transgenic mice have been used, in which the HCs express diphtheria toxin receptors followed by the *Pou4f3* promoter. In these mice, the administration of diphtheria toxin selectively ablates the HCs in vivo. In cases where hair cell death was induced at birth (postnatal day 0), limited spontaneous HC regeneration was observed. Subsequent fate-mapping experiments revealed that the neighboring SCs acquired the HC phenotype by following mitotic and direct transdifferentiation mechanisms

[49]. Other in vitro experiments provided evidence of limited spontaneous HC regeneration after aminoglycoside induced ototoxic damage in the neonatal mouse cochlea [45]. RNA seq analysis of neomycin-treated and untreated *Lgr5*+ SCs revealed major differences in expression of cell cycle genes, transcription factors, microRNAs, and cell signaling pathway genes that might be responsible for regulating proliferation and HC regeneration in the neonatal mouse cochlea [50, 51]. Detailed comparison through transcriptome expression profiles indicated that *Lgr5*+ SCs from the apex of the cochlea hold greater potential to regenerate HC as compared to basal *Lgr5*+ SCs in the neonatal mouse cochlea [52]. Also, inhibition of Notch signaling in the neonatal cochlea indirectly induces the Wnt signaling pathway; this allows the proliferation and mitotic regeneration of SCs in in vivo and in vitro models and the majority of the regenerated HCs are derived from *Lgr5*+ progenitors in the murine cochlea [53].

Similarly, the genetic reprogramming of *Sox2*+ SC by simultaneous upregulation of Wnt signaling, downregulation of Notch signaling, and overexpression of *Atoh1* promotes extensive proliferation of SCs and HC regeneration in the neonatal mouse cochlea [54]. In addition to the *Lgr5* gene, *Lgr6* is another Wnt downstream target gene that is expressed in the neonatal mouse cochlea [55]. On the basis of in vitro cultures and transcriptome analysis, *Lgr6*+ SCs exhibited greater proliferation and HCs regeneration capacity compared to the *Lgr5*+ SCs in the neonatal mouse cochlea [56]. Moreover, *Bmi1*, a member of polycomb protein family, also regulates the proliferation of *Lgr5*+ and other SCs. Deficiency of *Bmi1* significantly reduced the proliferation capacity of *Lgr5*+ progenitors in the neonatal mouse cochlea [57]. In cultured cochlear explants, the sonic hedgehog protein enhanced sphere generation, proliferation, and differentiation of *Lgr5*+ SCs. The overexpression of hedgehog signaling after neomycin injury leads to extensive SC proliferation and HC regeneration in cochlear explants [58]. Similarly, *Stat3* signaling also regulates HC differentiation during mouse inner ear development and the conditional deletion of *Stat3* inhibits HC differentiation in vivo and in vitro using a transgenic mouse model [59]. Collectively, these studies identify many different signaling pathways that might be used to induce proliferation and the HC regeneration in the cochlear sensory epithelium.

As compared to the cochlear sensory epithelium, HCs regeneration is more robust in the neonatal mouse utricle. Selective damage of HCs in the utricle results in activation of *Lgr5*+ among SCs located in the striolar region of the epithelium, which was associated with HCs regeneration through both mitotic and transdifferentiation pathways [60–62]. Also, the *Lgr5*+ SCs in the striolar region are more responsive to Wnt and Notch signaling compared to the other SCs in the striolar region [63]. The use of a combined approach to simultaneously inhibit the Notch signaling and overexpress the Wnt signaling pathway leads to the significantly greater SC proliferation and HC regeneration in normal and gentamicin-damaged postnatal mouse utricles [64]. By contrast, the mature mammalian utricle has shown limited regenerative potential after HCs damage [33, 35, 36]. Recent findings using a diphtheria toxin inducible transgenic mouse model illustrated the limited capacity of HC regeneration in the mature utricle. In this study, the number of HCs returning after HC damage was 17% greater than controls after 60 days of recovery. The new HCs were generated through direct transdifferentiation and had synaptic connections and mechanotransduction features [34, 65]. Unfortunately, in the neonatal cochlea, the newly regenerated HCs survive for some period, but most HCs degenerate gradually, for reasons that are poorly understood. Furthermore, in mammals, the SCs lose their natural regenerative potential with advancing age and this ability is rapidly reduced after the first postnatal week in the mature cochlea. Despite these limitations, these studies indicate that a subset of the SCs are present the inner ear and have the potential to regenerate new HCs. Future research breakthroughs may make it possible to develop therapeutic strategies to cure hearing loss.

2 Approaches to Regenerate HCs and SGNs

In recent years, there has been considerable effort made by hearing scientists to develop new therapeutic strategies to regenerate HCs and SGNs in order to restore normal hearing and provide relief to the millions of severe to profoundly impaired patients [66]. Because, cochlear HC loss increases the risk of subsequent SGN degeneration, it is important to preserve both the HCs and remaining SGNs in the cochlea. At present, the two most promising therapeutic approaches to regenerate cochlear HCs and SGNs involve gene therapy and stem cell therapy.

2.1 Gene Therapy

With gene therapy, the goal is to introduce the desired exogenous gene into the target cell in order to replace or fix the defective gene, thereby preventing or curing the disorder associated with that defective gene [67, 68]. Successful gene therapy in the targeted cells results in the long-term production of the desired protein in the cell. The small resident population of cells in the cochlea are immersed in perilymph and endolymph. These conditions allow for easy diffusion of vectors into the cochlea, making it a suitable target for gene therapy [69, 70].

Cochlear Gene Therapy Vehicles

The choice of vector, delivery method, and promoter has an influence on the transgene expression. The two most common strategies employed for gene therapy involve: (1) gene transfer via viruses (termed transduction) and (2) gene transfer via nonviral vectors (termed transfection). A number of different viral vectors have been used such as Adenovirus (Ad), Adeno-associated virus (AAV), synthetic AAV, herpes simplex virus, lentivirus, and vaccinia virus [71]. The nonviral vectors used in cochlear gene therapy include liposomes, plasmids, and synthetic cationic

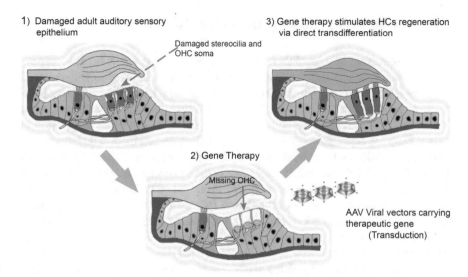

Fig. 2 Schematic diagram illustrating gene therapy via AAV vectors in the damaged mammals auditory sensory epithelium that stimulates the direct transdifferentiation of SCs into HCs. (1) Note damage to stereocilia and shrunken soma of OHC. (2) Note missing OHC and treatment of damaged cochlea with AAV gene therapy. (3) Gene therapy results in transdifferentiation of Deiters cells to OHC, resulting in the loss of SC beneath the regenerated OHC

polymers [72]. In general, AD virus vectors are the most widely used for human gene therapy to treat different human diseases [73]. However, AAV has been the principle viral vector used for gene therapy to cure SNHL (Fig. 2) because it is less toxic and immunogenic than Ad [74, 75]. The use of different promoters and sero-types affects the targeting capacity of AAV to different cell types and the expression level of of the transgenes [71]. AAV serotypes 1, 2, and 5 efficiently transduce HCs and SCs in cochlear explants of neonatal mice [76]. The use of AAV2/2 sero-types to efficiently pass through the round window membrane primarily transduces cells in the spiral limbus, SGNs, and spiral ligament [77, 78]. Moreover, in vivo findings show that the 12 different serotypes of AAV vectors, carrying a reporter gene, transduce into neonatal and adult mice cochlea targeting specific cell types and are expressed at different levels in the different targeted cells [74]. Recently, exome-associated AAV vectors (exo-AAV) have been found to be effective vehi-cles to deliver transgenes into cochlear and vestibular HCs both in vivo and in vitro [79]. A comparison of five different AAV vectors were examined in vivo and in vitro. Among them AAV2/Anc80L65-CMV vector was found to be the most promising vehicle for gene therapy in patients because of its high transduction rates in HCs [80].

In summary, most studies have focused on optimizing AAV vectors to success-fully replace genes into HCs and SCs in order to restore hearing function. Other viral and nonviral vectors are not discussed here because they are not as efficient at transferring genes of interest into HCs, SCs, or SGN in the cochlea.

2.2 Key Signaling Pathway Genes for HC Regeneration

As discussed above, cochlear SCs retain some regenerative potential because they are able to proliferate and differentiate into HCs. Therefore, in order to stimulate HC regeneration in adults, it is important to understand the primary gene signaling pathways responsible for initiating proliferation and differentiation. Cyclin-dependent kinase inhibitor (*p27kip1*) is one of the key regulators of cell cycle progression in SCs. *p27kip1* is expressed in SCs where it normally prevents these cells from undergoing cell division and proliferation during the postnatal period. The deliberate inhibition of the *p27kip1* gene in transgenic mouse models enables SCs to reenter the cell cycle thereby allowing SC proliferation to occur [81, 82]. Thus, inhibition of the *p27kip1* provides the basis for stimulating HCs regeneration in adult mammals. Besides, the ability of SCs to rapidly proliferate after *p27kip1* deletion, these newly divided cells disappear soon afterward via apoptosis. Other cell cycle regulatory genes such as retinoblastoma (*Rb*) [83], *p19 Ink4d* [84] and *Wnt* [85, 86] also regulate proliferation. The conditional deletion of *Rb* in HCs of newborn mice displayed that the HCs rapidly reenter into the cell cycle but unable to generate supernumerary HCs and eventually died [87]. Similarly, the selective ablation of *Rb* in postmitotic SCs (pillar and Deiter's cells) trigger the cell cycle reentry of SCs but only pillar cells showed the ability to proliferate and maintain their SC fate and survive for more than a week. However, these SCs also failed to generate new HCs [88]. Recently, the development of DN-CBRb mice (dominant-negative tetracycline-inducible *TetO-CB-myc6-Rb1*) to observe the transient and inducible dominant inhibition of the endogenous Retinoblastoma protein displayed the generation of supernumerary inner hair cells in the lower cochlear turns at P10 and P28 mice. This suggests that the targeted deletion of *Rb1* has an important role in stimulating HC regeneration in adult mammalian cochlea. *p19 Ink4d* is another cell cycle regulatory gene responsible for maintaining the quiescent state of cochlear cells in postnatal mice. The targeted deletion of *Ink4d* is sufficient to initiate the reentry of HCs into the cell cycle. However, these HCs rapidly undergo apoptosis resulting in a progressive hearing loss in postnatal mice [89]. The deliberate expression of *Wnt* gene both in vivo and in vitro during the postnatal period significantly promotes the SCs proliferation and generation of new HCs [46, 47, 90, 91]. Therefore, the cochlear SCs loose their ability to respond to *Wnt* with the advance age. Thus, a handful of genes have been identified that can stimulate cell proliferation in the inner ear. Controlled proliferation would seem to be an important requirement to replace SCs that differentiate or transdifferentiate into HC.

2.3 Atoh1-Based Gene Therapy

Another important regulator of inner ear development is Atonal homolog 1 (*Atoh1*, previously called *Math1* in mice). *Atoh1* is an essential gene required for HCs formation in the inner ear. Overexpression of *Atoh1* stimulates the production of extra

HCs in the postnatal rat inner ear. Recent gene therapeutic approaches with *Atoh1* seek to regenerate HCs direct transdifferentiation of SCs into HCs. But transdifferentiation has a downside, namely the depletion of critically important SCs. The first human clinical trial for hearing restoration "CGF166" using adenovirus vector encoding human Atonal homolog 1 (*Hath1*) is still underway since 2014. The preliminary findings regarding the effectiveness and feasibility of *Atoh1* gene in humans are unpublished yet [92].

The *Atoh1* gene plays a crucial role in the HC development and HC fate decision. *Atoh1* is an atonal basic helix–loop–helix transcription factor that participates in regulating HCs generation from SCs in the cochlea [93–95]. The targeted deletion of the *Atoh1* gene in a transgenic mouse model completely halted the development of the cochlear sensory epithelium and neighboring SCs [95]. In contrast, the overexpression of *Atoh1* transcription factor in postnatal rat cochlear explants promoted extra HCs-like cells in the cochlea [96]. Another, transcription factor *Pax2* (paired box 2), which is expressed during the early development of inner ear, promotes SC proliferation. Cotransfection of *Pax2* and *Atoh1* (*Math1*) in the neomycin-damaged organ of Corti triggered resident SCs to proliferate and differentiate into functional HCs in situ [97].

Moreover, a recent study found that combined overexpression of *Atoh1*, zinc finger transcription factor *GFI1* (required for HCs differentiation and survival) and *Pou4f3* (primarily required for the HC development in the inner ear) in a human fibroblast cell culture lead to the conversion of fibroblast cells into cells with HC-like features in vitro [98]. These results are particularly exciting because fibroblasts can be harvested from a patient, readily expanded because of their proliferative properties, thereby providing a pool offspring that could be used for laboratory testing or possibly transferred back into the patient's own ear. However, in vivo trials with this combinatorial gene therapy approach are still required. In cases of long-term deafness, one complication to this approach, as well as other approaches, is gradual loss of SCs that results in a flattened squamous epithelium.

The adenoviral-mediated overexpression of *Hath1*, human atonal homolog (similar to *Atoh1*) in postnatal rat cochlear explant cultures stimulates the robust production of new HCs following cochlear damage induced by aminoglycoside antibiotics. This work provides the basis for in vivo gene therapy work in the mature mammalian cochlea [99]. In the in vivo model, the deliberate insertion of the *Atoh1* gene into the cochlea of mature, deaf guinea pigs using an adenoviral vector stimulated the apparent formation of new HCs by transdifferentiation of SCs. The new HCs in the organ of Corti appeared immature, features consistent recent transdifferentiation. The *Atoh1*-driven production of new, immature-looking HC was associated with improved hearing thresholds in the drug-deafened guinea pigs. Electron microscopic examination of *Atoh1*-treated cochlea suggested transdifferentiation of SC into cells with HCs like features [100]. However, subsequent studies were unable to reproduce the same results in a mature, deaf guinea pig model.

Similarly, in utero *Atoh1* gene transfer using an expression plasmid generated supernumerary HCs in the mouse cochlea. These newly regenerated HCs displayed the characteristic features of normally developing HCs such as stereociliary bundles, attachment of neuronal processes, expression of the ribbon synapse marker *CtBP2*, and mechanoelectrical transduction [101]. However, one study on *Atoh1* gene therapy reported a significant increase in the number of newly formed HCs which expressed the HCs marker myosin7a. However, after a 3-week period, no improvement in hearing thresholds were observed in the mice after *Atoh1* transfection [102]. However, other studies in ototoxic-damaged mice observed some recovery of hearing thresholds after 2 months of *Atoh1* transfection [103, 104]. These studies indicated that it takes at least takes 3 weeks for SCs to convert into HCs and 2 months for hearing to recover after *Atoh1* gene insertion.

2.4   Inactivation of Cell Cycle Inhibitors Based Gene Therapy

The second approach to regenerate cochlear HCs is by inhibiting the cell cycle inhibitors *p27kip1* and *Rb* in SCs, thus allowing SCs to regain their proliferative capacity. The cyclin-dependent kinase inhibitor *p27kip1* is expressed during embryonic development of the organ of Corti. Its expression in otic epithelial progenitors interrupts cell division progression in these cells allowing for the differentiation of newborn cells into HCs and SCs from otic progenitors. Subsequently, the expression of *p27kip1* is only decreased in the HCs, while the SCs continue to express *p27kip1* in the mature cochlea. The knockdown of *p27kip1* expression in neonatal mice cochlear SCs results in regeneration of HCs via both mitotic and direct transdifferentiation mechanisms. However, after 2 weeks of postnatal life the resident SCs in the cochlear sensory epithelium, unable to downregulate *p27kip1* expression, thus lose their ability to return into the cell cycle [42]. The use of short hairpin RNA expressing vectors to knockdown the expression of *p27kip1* in the postnatal mouse cochlea causes cell cycle reentry of postnatal cochlear SCs. However, the activation of apoptotic pathway is also observed in some SC [105]. Similarly, the conditional deletion of *p27kip1* gene, in sensory HCs in early postnatal life of mice results in the proliferation and survival of these cells to adulthood without any hearing deficit [106]. Another recent approach to regenerate new HCs in the mature mouse cochlea after noise-induced damage is to combine the two genetic manipulations in SCs: *p27kip1* deletion and ectopic *Atoh1* expression [107]. Collectively, these studies suggested that the inhibition of cyclin-dependent kinase inhibitor is important to trigger HC regeneration in mammals.

Retinoblastoma (*Rb*) is another crucial negative regulator of cell cycle progression that controls the cell cycle exit of embryonic mammalian HCs. *Rb* expression in HCs and SCs is mainly required to maintain HCs in a quiescent state; manipulation of *Rb* cell cycle machinery disrupts the development of HCs. Conditional deletion of *Rb* in the inner ear of mice triggers cell cycle reentry and encourages the proliferation of HCs and SCs in the cochlea [88, 108, 109]. However, this manipulation also results in disturbances in the newly generated HCs such as promoting apoptosis and chromosomal abnormalities [108]. Although HC proliferation occurs in *Rb* knockout mice, the mice develop profound hearing loss due to the degeneration of organ of the Corti at 3 months of age. These results suggest that *Rb* gene expression may be essential for HC maturation and/or survival [109]. The targeted inactivation of *Rb* using the HC-specific inducible Cre system in cochlear HCs of newborn mice showed that approximately 40% of the HCs reentered the cell cycle by postnatal day 4. However, the HCs were unable to complete the process of cell division and the number of HC diminished rapidly causing profound deafness [110]. In conclusion, these studies indicate that complete inactivation of the *Rb* gene alone will give rise to functional HC. Thus, temporary *Rb* inactivation plus some other genetic manipulation may provide a useful strategy to promote successful HC regeneration and aid in recovery of hearing provided that the new HCs can make functional synapses with SGNs.

2.5   Gene Therapy for SGN Regeneration

SGNs, like HCs, also lack regeneration ability. Once SGNs are lost, it results in permanent hearing deficits [11, 111]. The successful use of prosthetic devices such as hearing aids and cochlear implants depends on having a large population of SGNs so that information can be relayed from the cochlea to the central auditory system. In the absence of SGNs, hearing aids and cochlear implants will be of no values to the hearing impaired hearing loss. However, the regeneration and replacement of SGNs are possible solutions to restore hearing in these patients [112]. Initially most of the work on SGN gene therapy focused on the preservation of residual SGNs using different neurotrophic factors such as neurotrophin 3 (NT3), brain-derived neurotrophic factor (BDNF), and bone morphogenetic protein 4 (BMP4) in order to support the survival of SGNs and their neurite outgrowth to the sensory HCs [113–116]. Later, different gene therapeutic strategies were tried to induce a neuronal phenotype in resident nonneural cells in the cochlea.

Ascl1 is a key neurogenic basic helix loop helix transcription factor that retains the capacity to transform different cell types into neuronal cells [117, 118]. The targeted overexpression of *Ascl1* alone or together with *Neurod1* in cochlear non-sensory epithelial cells stimulates the conversion of non-sensory cells into functional neurons in the embryonic and postnatal mouse cochlea. Also, the newly formed neurons displayed characteristics representative of neurons in terms of morphology, gene expression, and electrophysiology [118]. Although these transcription factors are able to convert these distant cell types into neuronal-like cell, the induced neuronal cells derived from non-sensory cells take longer to mature. Reserves of glial cells in Rosenthal's canal surrounding the SGNs represent an optimal cell type to convert into SGNs. The overexpression of *Ascl1* and neuron differentiation factor *Neurod1* in glial cells results in the generation of neuronal cells at high efficiency. The transcriptome analysis of these induced neuron cells revealed that the gene expression profile resembled those of resident SGNs such as the expression of neuronal markers *Prox1*, *Tubb3*, *Map2*, *Prph*, and *Snap25*. Moreover, when cocultured with the organ of Corti or cochlear nucleus, these induced neuron cells produced outgrowth toward the cochlear HCs and cochlear nucleus neurons [119, 120].

Another recent study focusing on neural preservation and protection used adeno-associated virus vector serotype 5 (AAV-5) as a gene therapy vector to induce high-level transgene expression of glial-derived neurotrophic factor (GDNF) in cochlear SGNs. The intracochlear injection of AAV-5 encoding neurotrophic factors may provide a suitable delivery method for hearing restoration and protection in the future [121]. GDNF is a potent nerve growth and survival factor, which is similar in nature as BDNF and NT3. GDNF served as a key survival factor for midbrain dopaminergic, spinal motor, cranial sensory, and sympathetic neurons [122–124] and it is also expressed in SGNs and the sensory epithelium during embryonic development as well as in the mature cochlea [125, 126]. The administration of GDNF into the cochlea has been reported to improve the survival of SGN and protect HCs damage from intense noise exposure in guinea pig [127–129]. Thus, GDNF may be an ideal candidate gene to determine the optimum transfection efficiency simultaneously for both HCs and SGNs.

3   Stem Cell Therapy

Stem cells, the master cells of the body, preserve their undifferentiated state and are capable of either directly differentiating into specialized cells or alternatively, follow the mitotic division path to develop more stem cells. Thus, stem cells are used to repair cellular damage and replace cell loss. Stem cell therapeutics is also sometimes referred to as regenerative medicine. Different types of stem cells are used in therapeutics such as the embryonic stem cells, adult stem cells, and inducible pluripotent stem cells (iPSCs) [130]. iPSC were first generated from skin fibroblasts by introducing only four transcription factors: *Sox2, c-Myc, Oct3/4,* and *Klf4* [131]. The major benefit of iPSCs in therapeutics is the lack of immunorejection because the iPSCs can be generated from the patient's own adult cells and later reintroduced into the same patient. This also reduces the ethical concerns associated with human embryonic stem cell therapy.

Stem cell therapy has been considered as a potential approach to restore missing HCs and restore hearing. Two possible stem cells could be used to treat deafness: endogenous stem cells or exogenous stem cells. Endogenous stem cell rehabilitation involves the activation of resident progenitor cells within the cochlear sensory epithelium with the goal of replacing damaged or missing HCs and to restore hearing. One major limitation to this approach is that endogenous progenitors in the inner ear are very limited in number hindering their ability to restore major forms of cochlear damage. The supply problem can be overcome with the unlimited supply of exogenous stem cells, an approach referred to as stem cell transplantation. Exogenous stem cells can be transplanted into the inner ear by several routes. One involves cells transplantation into scala tympani via round window or through a cochleostomy and then stimulating them to move through the basilar membrane or surrounding tissue into the organ of Corti where the HCs normally reside [132]. The second approach involves directly injecting the cells into scala media; this approach improves the chances that the cells will migrate into the HC region of the organ of Corti.

Many attempts have been made to stimulate exogenous stem cells to migrate into the organ of Corti and become HCs in vitro. Mouse embryonic stem cells were first used to generate HC-like cells in vitro using a stepwise differentiation strategy. Gene expression profiling and immunolabeling revealed that the differentiated cells displayed the comprehensive expression of hair cells specific markers observed via gene expression profile and immunostaining [133]. Others have successfully differentiated mouse pluripotent stem cells, embryonic stem cells, and iPSCs into HC-like cells in vitro. HC differentiation was accomplished by employing strategies that mimicked the fundamental principles of early embryonic and normal otic development. Using an in vitro approach, embryonic stem cells were placed on a feeder layer of chicken utricle stromal cells leading to the differentiation and maturation of HC-like cells. These cells had mechanosensing stereociliary structures on their surfaces similar to the mouse vestibular HCs and were responsive to the mechanical stimulation [134]. In this comprehensive report, different strategies were formulated using the feeder cell layers. One recent in vitro study found that the combined use of a feeder cell layer (ST2 stromal cell conditioned medium) and transfection of *Atoh1* efficiently induced HC-like cells from mouse embryonic stem cells [135]. In addition to a feeder cell layer, some other strategies using three-dimensional culture systems have been used to convert mouse embryonic stem cells into HCs-like cells, SCs, and neuronal cells that established synaptic connectivity with HCs [136]. Other crucial discoveries included the identification of transcriptional machinery that controls HC fate and differentiation. The combination of three transcription factors, *Gfi1*, *Pou4f3*, and *Atoh1*, and the expression of these factors in mouse embryonic stem cells directly stimulate genetic programming towards HCs. The newly formed HCs showed the HC-specific markers as confirmed by the transcriptome profiling and displayed the polarized membrane protrusions reminiscent of stereociliary bundles [137]. Moreover, a recent study defined a protocol to generate inner ear organoids. The first step involved deriving mouse embryonic stem cells from the blastocysts of a Pax2 fluorescent reporter mouse line. Then, these $Pax2^{EGFP/+}$ cells were place in a three-dimensional culture system to generate an inner ear organoid. During organoid development, increased expression of *Pax2* and activation of ERK occurred downstream of the FGF signaling pathway. The expression of *Pax2* was persistent throughout the generation of sensory HCs and the period when cochlear neurons formed synaptic connections with HCs in the organoids [138].

Few studies have employed human embryonic stem cells to generate HCs-like cells. In one case, human embryonic stem cells were programmed to differentiate under specific signals required for the specification of an early otic placode. Some otic progenitors differentiated into HCs-like cells expressing HC-specific marker and immature stereociliary bundles and possessed electrophysiological features characteristics of functional HCs. Other progenitors differentiated into neuronal cells expressing specific neuronal markers with electrophysiological properties characteristic of auditory neurons [139]. The formation of human otic progenitors cells depended on fibroblast growth factor signaling; the HCs-like cells displayed HC specific markers and rudimentary stereociliary bundles. However, these HCs failed to develop a fully mature HCs cytoarchitecture [140].

After the HC differentiation protocols were established, other researchers used these protocols to generate iPSCs from persons carrying *Myo7a* and *Myo15* mutations responsible for deafness. They used the new gene editing tool CRISPR/CAS 9 to genetically correct the *Myo7a* and *Myo15* mutations in the iPSCs. The corrected iPSCs were used to derive healthy HCs-like cells with proper organization of the stereociliary and cell body as well as normal functional properties [141, 142]. Also, inner ear organoids have been generated from human embryonic stem cells using a three-dimensional culture system. This human inner ear organoid contained HCs, SCs, and neuronal cells with the proper cytoarchitecture in the inner ear sensory epithelium [143]. The organoid culture system facilitates the study of human inner ear development and provides an in vitro model that can be used to study inner ear damage and new therapeutic approaches. More recently, human urinary cells collected from healthy individuals have been reprogrammed into iPSCs. These iPSC were stimulated to differentiate into otic epithelial cells and HC-like cells and were cocultured with SGNs. The induced HC-like cells had many of the morphological and electrophysiological characteristics of native HCs. These newly generated HC-like cells formed synaptic connections with SGNs in coculture. However, when these cultured iPSC were transplanted to the inner ear, very few of them migrated to the native HCs sites in the organ of Corti and few formed synaptic connections with resident SGNs [144].

In vitro studies of both mouse and human embryonic stem cells and iPSCs have shown that under controlled culture conditions, iPSC attain the desired cell fate in the form of HCs and neuronal cells [145, 146]. When these progenitors are introduced at the site of epithelial injury in the inner ear, some transplanted cells can integrate into the developing inner ear and express HC specific markers and hair bundles in the cochlear or vestibular sensory epithelia in vivo [133]. In contast, very few studies have reported integration of these newly generated HCs into the mammalian inner ear. Therefore, at the present time, the delivery of stem cells aimed at generating functional HCs at the site of damage in the mammalian inner ear remains a significant challenge with an uncertain future [147, 148]. Few transplanted cells form the desired cell fate (e.g, HCs, SCs, neuronal and glial cells) even after several weeks of transplantation. One reason for the lack of in vivo success could be that the in vivo microenvironment in the mammalian cochlea is completely different from in vitro culture conditions. Another in vivo complication is failure of the transplanted cells to correctly home to and integrate into the damaged cochlear regions where they are needed and then become functional HCs. A third major challenge is the proper HCs integration and orientation within the elegantly sculted cochlea where every HC, SC, and neuron has a specific location and unique orientation.

4   Conclusion

Formulation of different strategies to regenerate HC in mammals is one of the most lofty and clinically important goals of hearing research. While there have been major advances in the regeneration of HCs, SCs, and SGN in the past three decades using animal models, major challenges lie ahead in translating the in vitro advances to allow for clinical applications of exogenous stem cells transplantation and gene therapy approaches aimed at regenerating HCs in the mammalian inner ear. Several challenging questions need to be addressed before implementing these therapies in humans such as the safety of viral vectors used in clinical trials, the potential adverse effects to the patients, proper and controlled exogenous stem cell growth at the site of cell implantation. Also, the high cost of preparing stem cells and developing viral vector gene therapy may initially make it an unreachable goal for the majority of hearing loss patients. In addition, a major question is whether these therapies in humans or animal models are able to restore hearing to a level where normal speech or musical perception is achieved. Despite these concerns, exogenous stem cells and gene therapy represent exciting and promising approaches to elucidate the conditions needed to intiate HC regeneration in the mature mammalian cochlea, something that nature has successfully achieved in many aves, amphibians, and other nonmamallian species.

References

1. Hudspeth A (2014) Integrating the active process of hair cells with cochlear function. Nat Rev Neurosci 15(9):600
2. Fettiplace R, Hackney CM (2006) The sensory and motor roles of auditory hair cells. Nat Rev Neurosci 7(1):19
3. He DZ, Jia S, Dallos P (2004) Mechanoelectrical transduction of adult outer hair cells studied in a gerbil hemicochlea. Nature 429(6993):766
4. Avan P, Büki B, Petit C (2013) Auditory distortions: origins and functions. Physiol Rev 93(4):1563–1619
5. Moser T, Starr A (2016) Auditory neuropathy—neural and synaptic mechanisms. Nat Rev Neurol 12(3):135–149. https://doi.org/10.1038/nrneurol.2016.10
6. Mcgill TJ, Schuknecht HF (1976) Human cochlear changes in noise induced hearing loss. Laryngoscope 86(9):1293–1302
7. Nadol JB Jr, Burgess BJ, Gantz BJ, Coker NJ, Ketten DR, Kos I, Roland JT Jr, Shiao JY, Eddington DK, Montandon P (2001) Histopathology of cochlear implants in humans. Ann Otol Rhinol Laryngol 110(9):883–891
8. McHaney V, Thibadoux G, Hayes F, Green A (1983) Hearing loss in children receiving cis-platin chemotherapy. J Pediatr 102(2):314–317
9. Rybak LP (1986) Drug ototoxicity. Annu Rev Pharmacol Toxicol 26(1):79–99
10. Waqas M, Gao S, Ali MK, Ma Y, Li W (2018) Inner ear hair cell protection in mammals against the noise-induced cochlear damage. Neural Plast 2018:3170801
11. Kujawa SG, Liberman MC (2009) Adding insult to injury: cochlear nerve degeneration after "temporary" noise-induced hearing loss. J Neurosci 29(45):14077–14085
12. Kujawa SG, Liberman MC (2015) Synaptopathy in the noise-exposed and aging cochlea: primary neural degeneration in acquired sensorineural hearing loss. Hear Res 330:191–199
13. WHO (20 Mar 2019) Deafness and hearing loss report. https://www.who.int/en/news-room/fact-sheets/detail/deafness-and-hearing-loss
14. Tseng CC, Hu LY, Liu ME, Yang AC, Shen CC, Tsai SJ (2016) Risk of depressive disorders following sudden sensorineural hearing loss: a nationwide population-based retrospective cohort study. J Affect Disord 197:94–99. https://doi.org/10.1016/j.jad.2016.03.020

15. Homans NC, Metselaar RM, Dingemanse JG, van der Schroeff MP, Brocaar MP, Wieringa MH, Baatenburg de Jong RJ, Hofman A, Goedegebure A (2017) Prevalence of age-related hearing loss, including sex differences, in older adults in a large cohort study. Laryngoscope 127(3):725–730. https://doi.org/10.1002/lary.26150
16. Goman AM, Reed NS, Lin FR (2017) Addressing estimated hearing loss in adults in 2060. JAMA Otolaryngol Head Neck Surg 143(7):733–734. https://doi.org/10.1001/jamaoto.2016.4642
17. Sun LW, Johnson RD, Langlo CS, Cooper RF, Razeen MM, Russillo MC, Dubra A, Connor TB Jr, Han DP, Pennesi ME, Kay CN, Weinberg DV, Stepien KE, Carroll J (2016) Assessing photoreceptor structure in retinitis Pigmentosa and usher syndrome. Invest Ophthalmol Vis Sci 57(6):2428–2442. https://doi.org/10.1167/iovs.15-18246
18. Tomblin JB, Harrison M, Ambrose SE, Walker EA, Oleson JJ, Moeller MP (2015) Language outcomes in young children with mild to severe hearing loss. Ear Hear 36(1):76S
19. Bush ALH, Lister JJ, Lin FR, Betz J, Edwards JD (2015) Peripheral hearing and cognition: evidence from the staying keen in later life (SKILL) study. Ear Hear 36(4):395
20. Lin FR, Metter EJ, O'Brien RJ, Resnick SM, Zonderman AB, Ferrucci L (2011) Hearing loss and incident dementia. Arch Neurol 68(2):214–220
21. Smith SL, Pichora-Fuller MK (2015) Associations between speech understanding and auditory and visual tests of verbal working memory: effects of linguistic complexity, task, age, and hearing loss. Front Psychol 6:1394
22. Sprinzl G, Riechelmann H (2010) Current trends in treating hearing loss in elderly people: a review of the technology and treatment options—a mini-review. Gerontology 56(3):351–358
23. Hunter CR, Kronenberger WG, Castellanos I, Pisoni DB (2017) Early postimplant speech perception and language skills predict long-term language and neurocognitive outcomes following pediatric cochlear implantation. J Speech Lang Hear Res 60(8):2321–2336. https://doi.org/10.1044/2017_jslhr-h-16-0152
24. Dowell RC, Dettman SJ, Blamey PJ, Barker EJ, Clark GM (2002) Speech perception in children using cochlear implants: prediction of long-term outcomes. Cochlear Implants Int 3(1):1–18
25. Huang J, Sheffield B, Lin P, Zeng F-G (2017) Electro-tactile stimulation enhances cochlear implant speech recognition in noise. Sci Rep 7(1):2196
26. Bruns L, Mürbe D, Hahne A (2016) Understanding music with cochlear implants. Sci Rep 6:32026
27. Corwin JT, Cotanche DA (1988) Regeneration of sensory hair cells after acoustic trauma. Science 240(4860):1772–1774
28. Ryals BM, Rubel EW (1988) Hair cell regeneration after acoustic trauma in adult Coturnix quail. Science 240(4860):1774–1776
29. Cruz RM, Lambert PR, Rubel EW (1987) Light microscopic evidence of hair cell regeneration after gentamicin toxicity in chick cochlea. Arch Otolaryngol Head Neck Surg 113(10):1058–1062
30. Weisleder P, Rubel EW (1992) Hair cell regeneration in the avian vestibular epithelium. Exp Neurol 115(1):2–6
31. Stone JS, Cotanche DA (2007) Hair cell regeneration in the avian auditory epithelium. Int J Dev Biol 51(6–7):633–647. https://doi.org/10.1387/ijdb.072408js
32. Soucek S, Michaels L, Frohlich A (1986) Evidence for hair cell degeneration as the primary lesion in hearing loss of the elderly. J Otolaryngol 15(3):175–183
33. Forge A, Li L, Nevill G (1998) Hair cell recovery in the vestibular sensory epithelia of mature guinea pigs. J Comp Neurol 397(1):69–88
34. Golub JS, Tong L, Ngyuen TB, Hume CR, Palmiter RD, Rubel EW, Stone JS (2012) Hair cell replacement in adult mouse utricles after targeted ablation of hair cells with diphtheria toxin. J Neurosci 32(43):15093–15105. https://doi.org/10.1523/jneurosci.1709-12.2012
35. Kawamoto K, Izumikawa M, Beyer LA, Atkin GM, Raphael Y (2009) Spontaneous hair cell regeneration in the mouse utricle following gentamicin ototoxicity. Hear Res 247(1):17–26. https://doi.org/10.1016/j.heares.2008.08.010

36. Lin V, Golub JS, Nguyen TB, Hume CR, Oesterle EC, Stone JS (2011) Inhibition of Notch activity promotes nonmitotic regeneration of hair cells in the adult mouse utricles. J Neurosci 31(43):15329–15339. https://doi.org/10.1523/jneurosci.2057-11.2011

37. Tanyeri H, Lopez I, Honrubia V (1995) Histological evidence for hair cell regeneration after ototoxic cell destruction with local application of gentamicin in the chinchilla crista ampullaris. Hear Res 89(1–2):194–202

38. Sinkkonen ST, Chai R, Jan TA, Hartman BH, Laske RD, Gahlen F, Sinkkonen W, Cheng AG, Oshima K, Heller S (2011) Intrinsic regenerative potential of murine cochlear supporting cells. Sci Rep 1:26. https://doi.org/10.1038/srep00026

39. Malgrange B, Belachew S, Thiry M, Nguyen L, Rogister B, Alvarez ML, Rigo JM, Van De Water TR, Moonen G, Lefebvre PP (2002) Proliferative generation of mammalian auditory hair cells in culture. Mech Dev 112(1–2):79–88

40. Oshima K, Grimm CM, Corrales CE, Senn P, Martinez Monedero R, Geleoc GS, Edge A, Holt JR, Heller S (2007) Differential distribution of stem cells in the auditory and vestibular organs of the inner ear. J Assoc Res Otolaryngol 8(1):18–31. https://doi.org/10.1007/s10162-006-0058-3

41. Oshima K, Senn P, Heller S (2009) Isolation of sphere-forming stem cells from the mouse inner ear. Methods Mol Biol 493:141–162. https://doi.org/10.1007/978-1-59745-523-7_9

42. White PM, Doetzlhofer A, Lee YS, Groves AK, Segil N (2006) Mammalian cochlear supporting cells can divide and trans-differentiate into hair cells. Nature 441(7096):984–987. https://doi.org/10.1038/nature04849

43. Li H, Liu H, Heller S (2003) Pluripotent stem cells from the adult mouse inner ear. Nat Med 9(10):1293–1299. https://doi.org/10.1038/nm925

44. Chai R, Xia A, Wang T, Jan TA, Hayashi T, Bermingham-McDonogh O, Cheng AG (2011) Dynamic expression of Lgr5, a Wnt target gene, in the developing and mature mouse cochlea. J Assoc Res Otolaryngol 12(4):455–469. https://doi.org/10.1007/s10162-011-0267-2

45. Bramhall NF, Shi F, Arnold K, Hochedlinger K, Edge AS (2014) Lgr5-positive supporting cells generate new hair cells in the postnatal cochlea. Stem Cell Rep 2(3):311–322. https://doi.org/10.1016/j.stemcr.2014.01.008

46. Chai R, Kuo B, Wang T, Liaw EJ, Xia A, Jan TA, Liu Z, Taketo MM, Oghalai JS, Nusse R (2012) Wnt signaling induces proliferation of sensory precursors in the postnatal mouse cochlea. Proc Natl Acad Sci 109(21):8167–8172

47. Shi F, Kempfle JS, Edge AS (2012) Wnt-responsive Lgr5-expressing stem cells are hair cell progenitors in the cochlea. J Neurosci 32(28):9639–9648. https://doi.org/10.1523/jneurosci.1064-12.2012

48. Atkinson PJ, Kim GS, Cheng AG (2018) Direct cellular reprogramming and inner ear regeneration. Expert Opin Biol Ther 19:129. https://doi.org/10.1080/14712598.2019.1564035

49. Cox BC, Chai R, Lenoir A, Liu Z, Zhang L, Nguyen D-H, Chalasani K, Steigelman KA, Fang J, Cheng AG (2014) Spontaneous hair cell regeneration in the neonatal mouse cochlea in vivo. Development (Cambridge, England) 141(4):816–829

50. Zhang S, Zhang Y, Yu P, Hu Y, Zhou H, Guo L, Xu X, Zhu X, Waqas M, Qi J, Zhang X, Liu Y, Chen F, Tang M, Qian X, Shi H, Gao X, Chai R (2017) Characterization of Lgr5+ progenitor cell transcriptomes after neomycin injury in the neonatal mouse cochlea. Front Mol Neurosci 10:213. https://doi.org/10.3389/fnmol.2017.00213

51. Cheng C, Guo L, Lu L, Xu X, Zhang S, Gao J, Waqas M, Zhu C, Chen Y, Zhang X, Xuan C, Gao X, Tang M, Chen F, Shi H, Li H, Chai R (2017) Characterization of the transcriptomes of Lgr5+ hair cell progenitors and Lgr5− supporting cells in the mouse cochlea. Front Mol Neurosci 10:122. https://doi.org/10.3389/fnmol.2017.00122

52. Waqas M, Guo L, Zhang S, Chen Y, Zhang X, Wang L, Tang M, Shi H, Bird PI, Li H, Chai R (2016) Characterization of Lgr5+ progenitor cell transcriptomes in the apical and basal turns of the mouse cochlea. Oncotarget 7(27):41123–41141. https://doi.org/10.18632/oncotarget.8636

53. Li W, Wu J, Yang J, Sun S, Chai R, Chen ZY, Li H (2015) Notch inhibition induces mitoti-cally generated hair cells in mammalian cochleae via activating the Wnt pathway. Proc Natl Acad Sci U S A 112(1):166–171. https://doi.org/10.1073/pnas.1415901112
54. Ni W, Lin C, Guo L, Wu J, Chen Y, Chai R, Li W, Li H (2016) Extensive supporting cell prolifer-ation and mitotic hair cell generation by in vivo genetic reprogramming in the neonatal mouse cochlea. J Neurosci 36(33):8734–8745. https://doi.org/10.1523/jneurosci.0060-16.2016
55. Zhang Y, Chen Y, Ni W, Guo L, Lu X, Liu L, Li W, Sun S, Wang L, Li H (2015) Dynamic expression of Lgr6 in the developing and mature mouse cochlea. Front Cell Neurosci 9:165. https://doi.org/10.3389/fncel.2015.00165
56. Zhang Y, Guo L, Lu X, Cheng C, Sun S, Li W, Zhao L, Lai C, Zhang S, Yu C, Tang M, Chen Y, Chai R, Li H (2018) Characterization of Lgr6+ cells as an enriched population of hair cell progenitors compared to Lgr5+ cells for hair cell generation in the neonatal mouse cochlea. Front Mol Neurosci 11:147. https://doi.org/10.3389/fnmol.2018.00147
57. Lu X, Sun S, Qi J, Li W, Liu L, Zhang Y, Chen Y, Zhang S, Wang L, Miao D, Chai R, Li H (2016) Bmi1 regulates the proliferation of cochlear supporting cells via the canonical Wnt signaling pathway. Mol Neurobiol 54:1326. https://doi.org/10.1007/s12035-016-9686-8
58. Chen Y, Lu X, Guo L, Ni W, Zhang Y, Zhao L, Wu L, Sun S, Zhang S, Tang M, Li W, Chai R, Li H (2017) Hedgehog signaling promotes the proliferation and subsequent hair cell forma-tion of progenitor cells in the neonatal mouse cochlea. Front Mol Neurosci 10:426. https://doi.org/10.3389/fnmol.2017.00426
59. Chen Q, Quan Y, Wang N, Xie C, Ji Z, He H, Chai R, Li H, Yin S, Chin YE, Wei X, Gao WQ (2017) Inactivation of STAT3 signaling impairs hair cell differentiation in the developing mouse cochlea. Stem Cell Rep 9(1):231–246. https://doi.org/10.1016/j.stemcr.2017.05.031
60. Li W, You D, Chen Y, Chai R, Li H (2016) Regeneration of hair cells in the mammalian ves-tibular system. Front Med 10(2):143–151. https://doi.org/10.1007/s11684-016-0451-1
61. Wang T, Chai R, Kim GS, Pham N, Jansson L, Nguyen DH, Kuo B, May LA, Zuo J, Cunningham LL, Cheng AG (2015) Lgr5+ cells regenerate hair cells via proliferation and direct transdifferentiation in damaged neonatal mouse utricle. Nat Commun 6:6613. https://doi.org/10.1038/ncomms7613
62. Burns JC, Cox BC, Thiede BR, Zuo J, Corwin JT (2012) In vivo proliferative regeneration of balance hair cells in newborn mice. J Neurosci 32(19):6570–6577. https://doi.org/10.1523/jneurosci.6274-11.2012
63. You D, Guo L, Li W, Sun S, Chen Y, Chai R, Li H (2018) Characterization of Wnt and Notch-responsive Lgr5+ hair cell progenitors in the striolar region of the neonatal mouse utricle. Front Mol Neurosci 11:137. https://doi.org/10.3389/fnmol.2018.00137
64. Wu J, Li W, Lin C, Chen Y, Cheng C, Sun S, Tang M, Chai R, Li H (2016) Co-regulation of the Notch and Wnt signaling pathways promotes supporting cell proliferation and hair cell regeneration in mouse utricles. Sci Rep 6:29418. https://doi.org/10.1038/srep29418
65. Bucks SA, Cox BC, Vlosich BA, Manning JP, Nguyen TB, Stone JS (2017) Supporting cells remove and replace sensory receptor hair cells in a balance organ of adult mice. Elife 6:e18128. https://doi.org/10.7554/eLife.18128
66. Lee MY, Park YH (2018) Potential of gene and cell therapy for inner ear hair cells. Biomed Res Int 2018:8137614. https://doi.org/10.1155/2018/8137614
67. Mulligan RC (1993) The basic science of gene therapy. Science 260(5110):926–932
68. Wang J, Puel JL (2018) Toward cochlear therapies. Physiol Rev 98(4):2477–2522. https://doi.org/10.1152/physrev.00053.2017
69. Ishimoto S, Kawamoto K, Kanzaki S, Raphael Y (2002) Gene transfer into supporting cells of the organ of Corti. Hear Res 173(1–2):187–197
70. Kanzaki S (2018) Gene delivery into the inner ear and its clinical implications for hearing and balance. Molecules 23(10):e2507. https://doi.org/10.3390/molecules23102507
71. Sacheli R, Delacroix L, Vandenackerveken P, Nguyen L, Malgrange B (2013) Gene trans-fer in inner ear cells: a challenging race. Gene Ther 20(3):237–247. https://doi.org/10.1038/gt.2012.51

72. Holley MC (2002) Application of new biological approaches to stimulate sensory repair and protection. Br Med Bull 63:157–169
73. Lim ST, Airavaara M, Harvey BK (2010) Viral vectors for neurotrophic factor delivery: a gene therapy approach for neurodegenerative diseases of the CNS. Pharmacol Res 61(1):14–26. https://doi.org/10.1016/j.phrs.2009.10.002
74. Shu Y, Tao Y, Wang Z, Tang Y, Li H, Dai P, Gao G, Chen ZY (2016) Identification of adeno-associated viral vectors that target neonatal and adult mammalian inner ear cell subtypes. Hum Gene Ther 27(9):687–699. https://doi.org/10.1089/hum.2016.053
75. Landegger LD, Pan B, Askew C, Wassmer SJ, Gluck SD, Galvin A, Taylor R, Forge A, Stankovic KM, Holt JR, Vandenberghe LH (2017) A synthetic AAV vector enables safe and efficient gene transfer to the mammalian inner ear. Nat Biotechnol 35(3):280–284. https://doi.org/10.1038/nbt.3781
76. Stone IM, Lurie DI, Kelley MW, Poulsen DJ (2005) Adeno-associated virus-mediated gene transfer to hair cells and support cells of the murine cochlea. Mol Ther 11(6):843–848. https://doi.org/10.1016/j.ymthe.2005.02.005
77. Li Duan M, Bordet T, Mezzina M, Kahn A, Ulfendahl M (2002) Adenoviral and adeno-associated viral vector mediated gene transfer in the Guinea pig cochlea. Neuroreport 13(10):1295–1299
78. Luebke AE, Foster PK, Muller CD, Peel AL (2001) Cochlear function and transgene expression in the guinea pig cochlea, using adenovirus- and adeno-associated virus-directed gene transfer. Hum Gene Ther 12(7):773–781. https://doi.org/10.1089/104303401750148702
79. Gyorgy B, Sage C, Indzhykulian AA, Scheffer DI, Brisson AR, Tan S, Wu X, Volak A, Mu D, Tamvakologos PI, Li Y, Fitzpatrick Z, Ericsson M, Breakefield XO, Corey DP, Maguire CA (2017) Rescue of hearing by gene delivery to inner-ear hair cells using exosome-associated AAV. Mol Ther 25(2):379–391. https://doi.org/10.1016/j.ymthe.2016.12.010
80. Gu X, Chai R, Guo L, Dong B, Li W, Shu Y, Huang X, Li H (2019) Transduction of adeno-associated virus vectors targeting hair cells and supporting cells in the neonatal mouse cochlea. Front Cell Neurosci 13:8. https://doi.org/10.3389/fncel.2019.00008
81. Oesterle EC, Chien WM, Campbell S, Nellimarla P, Fero ML (2011) p27(Kip1) is required to maintain proliferative quiescence in the adult cochlea and pituitary. Cell Cycle 10(8):1237–1248. https://doi.org/10.4161/cc.10.8.15301
82. Lowenheim H, Furness DN, Kil J, Zinn C, Gultig K, Fero ML, Frost D, Gummer AW, Roberts JM, Rubel EW, Hackney CM, Zenner HP (1999) Gene disruption of p27(Kip1) allows cell proliferation in the postnatal and adult organ of corti. Proc Natl Acad Sci U S A 96(7):4084–4088
83. Sage C, Huang M, Karimi K, Gutierrez G, Vollrath MA, Zhang DS, Garcia-Anoveros J, Hinds PW, Corwin JT, Corey DP, Chen ZY (2005) Proliferation of functional hair cells in vivo in the absence of the retinoblastoma protein. Science 307(5712):1114–1118. https://doi.org/10.1126/science.1106642
84. Cunningham JJ, Levine EM, Zindy F, Goloubeva O, Roussel MF, Smeyne RJ (2002) The cyclin-dependent kinase inhibitors p19(Ink4d) and p27(Kip1) are coexpressed in select retinal cells and act cooperatively to control cell cycle exit. Mol Cell Neurosci 19(3):359–374. https://doi.org/10.1006/mcne.2001.1090
85. Jansson L, Kim GS, Cheng AG (2015) Making sense of Wnt signaling-linking hair cell regeneration to development. Front Cell Neurosci 9:66. https://doi.org/10.3389/fncel.2015.00066
86. Waqas M, Zhang S, He Z, Tang M, Chai R (2016) Role of Wnt and Notch signaling in regulating hair cell regeneration in the cochlea. Front Med 10(3):237–249
87. Weber T, Corbett MK, Chow LM, Valentine MB, Baker SJ, Zuo J (2008) Rapid cell-cycle reentry and cell death after acute inactivation of the retinoblastoma gene product in postnatal cochlear hair cells. Proc Natl Acad Sci U S A 105(2):781–785. https://doi.org/10.1073/pnas.0708061105

88. Yu Y, Weber T, Yamashita T, Liu Z, Valentine MB, Cox BC, Zuo J (2010) In vivo proliferation of postmitotic cochlear supporting cells by acute ablation of the retinoblastoma protein in neonatal mice. J Neurosci 30(17):5927–5936. https://doi.org/10.1523/jneurosci.5989-09.2010

89. Chen P, Zindy F, Abdala C, Liu F, Li X, Roussel MF, Segil N (2003) Progressive hearing loss in mice lacking the cyclin-dependent kinase inhibitor Ink4d. Nat Cell Biol 5(5):422–426. https://doi.org/10.1038/ncb976

90. Samarajeewa A, Lenz DR, Xie L, Chiang H, Kirchner R, Mulvaney JF, Edge ASB, Dabdoub A (2018) Transcriptional response to Wnt activation regulates the regenerative capacity of the mammalian cochlea. Development 145(23):166579. https://doi.org/10.1242/dev.166579

91. Shi F, Hu L, Edge AS (2013) Generation of hair cells in neonatal mice by beta-catenin overexpression in Lgr5-positive cochlear progenitors. Proc Natl Acad Sci U S A 110(34):13851–13856. https://doi.org/10.1073/pnas.1219952110

92. NIH CTUNLoM (2019) Safety, tolerability and efficacy for CGF166 in patients with unilateral or bilateral severe-to-profound hearing loss. NIH. https://clinicaltrials.gov/ct2/show/NCT02132130

93. Zhong C, Fu Y, Pan W, Yu J, Wang J (2019) Atoh1 and other related key regulators in the development of auditory sensory epithelium in the mammalian inner ear: function and interplay. Dev Biol 446(2):133–141. https://doi.org/10.1016/j.ydbio.2018.12.025

94. Bermingham NA, Hassan BA, Price SD, Vollrath MA, Ben-Arie N, Eatock RA, Bellen HJ, Lysakowski A, Zoghbi HY (1999) Math1: an essential gene for the generation of inner ear hair cells. Science 284(5421):1837–1841

95. Woods C, Montcouquiol M, Kelley MW (2004) Math1 regulates development of the sensory epithelium in the mammalian cochlea. Nat Neurosci 7(12):1310–1318. https://doi.org/10.1038/nn1349

96. Zheng JL, Gao WQ (2000) Overexpression of Math1 induces robust production of extra hair cells in postnatal rat inner ears. Nat Neurosci 3(6):580–586. https://doi.org/10.1038/75753

97. Chen Y, Yu H, Zhang Y, Li W, Lu N, Ni W, He Y, Li J, Sun S, Wang Z, Li H (2013) Cotransfection of Pax2 and Math1 promote in situ cochlear hair cell regeneration after neomycin insult. Sci Rep 3:2996. https://doi.org/10.1038/srep02996

98. Duran Alonso MB, Lopez Hernandez I, de la Fuente MA, Garcia-Sancho J, Giraldez F, Schimmang T (2018) Transcription factor induced conversion of human fibroblasts towards the hair cell lineage. PLoS One 13(7):e0200210. https://doi.org/10.1371/journal.pone.0200210

99. Shou J, Zheng JL, Gao WQ (2003) Robust generation of new hair cells in the mature mammalian inner ear by adenoviral expression of Hath1. Mol Cell Neurosci 23(2):169–179

100. Izumikawa M, Minoda R, Kawamoto K, Abrashkin KA, Swiderski DL, Dolan DF, Brough DE, Raphael Y (2005) Auditory hair cell replacement and hearing improvement by Atoh1 gene therapy in deaf mammals. Nat Med 11(3):271–276. https://doi.org/10.1038/nm1193

101. Gubbels SP, Woessner DW, Mitchell JC, Ricci AJ, Brigande JV (2008) Functional auditory hair cells produced in the mammalian cochlea by in utero gene transfer. Nature 455(7212):537–541. https://doi.org/10.1038/nature07265

102. Atkinson PJ, Wise AK, Flynn BO, Nayagam BA, Richardson RT (2014) Hair cell regeneration after ATOH1 gene therapy in the cochlea of profoundly deaf adult guinea pigs. PLoS One 9(7):e102077. https://doi.org/10.1371/journal.pone.0102077

103. Kraft S, Hsu C, Brough DE, Staecker H (2013) Atoh1 induces auditory hair cell recovery in mice after ototoxic injury. Laryngoscope 123(4):992–999. https://doi.org/10.1002/lary.22171

104. Liu Z, Dearman JA, Cox BC, Walters BJ, Zhang L, Ayrault O, Zindy F, Gan L, Roussel MF, Zuo J (2012) Age-dependent in vivo conversion of mouse cochlear pillar and Deiters' cells to immature hair cells by Atoh1 ectopic expression. J Neurosci 32(19):6600–6610. https://doi.org/10.1523/jneurosci.0818-12.2012

105. Ono K, Nakagawa T, Kojima K, Matsumoto M, Kawauchi T, Hoshino M, Ito J (2009) Silencing p27 reverses post-mitotic state of supporting cells in neonatal mouse cochleae. Mol Cell Neurosci 42(4):391–398. https://doi.org/10.1016/j.mcn.2009.08.011

106. Walters BJ, Liu Z, Crabtree M, Coak E, Cox BC, Zuo J (2014) Auditory hair cell-specific deletion of p27Kip1 in postnatal mice promotes cell-autonomous generation of new hair cells and normal hearing. J Neurosci 34(47):15751–15763. https://doi.org/10.1523/jneurosci.3200-14.2014

107. Walters BJ, Coak E, Dearman J, Bailey G, Yamashita T, Kuo B, Zuo J (2017) In vivo interplay between p27(Kip1), GATA3, ATOH1, and POU4F3 converts non-sensory cells to hair cells in adult mice. Cell Rep 19(2):307–320. https://doi.org/10.1016/j.celrep.2017.03.044

108. Mantela J, Jiang Z, Ylikoski J, Fritzsch B, Zacksenhaus E, Pirvola U (2005) The retino-blastoma gene pathway regulates the postmitotic state of hair cells of the mouse inner ear. Development (Cambridge, England) 132(10):2377–2388. https://doi.org/10.1242/dev.01834

109. Sage C, Huang M, Vollrath MA, Brown MC, Hinds PW, Corey DP, Vetter DE, Chen ZY (2006) Essential role of retinoblastoma protein in mammalian hair cell development and hear-ing. Proc Natl Acad Sci U S A 103(19):7345–7350. https://doi.org/10.1073/pnas.0510631103

110. Wemeau JL, Kopp P (2017) Pendred syndrome. Best Pract Res Clin Endocrinol Metab 31(2):213–224. https://doi.org/10.1016/j.beem.2017.04.011

111. White JA, Burgess BJ, Hall RD, Nadol JB (2000) Pattern of degeneration of the spiral gan-glion cell and its processes in the C57BL/6J mouse. Hear Res 141(1–2):12–18

112. Dabdoub A, Nishimura K (2017) Cochlear implants meet regenerative biology: state of the science and future research directions. Otol Neurotol 38(8):e232–e236. https://doi.org/10.1097/mao.0000000000001407

113. Shibata SB, Cortez SR, Beyer LA, Wiler JA, Di Polo A, Pfingst BE, Raphael Y (2010) Transgenic BDNF induces nerve fiber regrowth into the auditory epithelium in deaf cochleae. Exp Neurol 223(2):464–472. https://doi.org/10.1016/j.expneurol.2010.01.011

114. Suzuki J, Corfas G, Liberman MC (2016) Round-window delivery of neurotrophin 3 regener-ates cochlear synapses after acoustic overexposure. Sci Rep 6:24907. https://doi.org/10.1038/srep24907

115. Wise AK, Tu T, Atkinson PJ, Flynn BO, Sgro BE, Hume C, O'Leary SJ, Shepherd RK, Richardson RT (2011) The effect of deafness duration on neurotrophin gene therapy for spiral ganglion neuron protection. Hear Res 278(1–2):69–76. https://doi.org/10.1016/j.heares.2011.04.010

116. Waqas M, Sun S, Xuan C, Fang Q, Zhang X, Islam IU, Qi J, Zhang S, Gao X, Tang M, Shi H, Li H, Chai R (2017) Bone morphogenetic protein 4 promotes the survival and pre-serves the structure of flow-sorted Bhlhb5+ cochlear spiral ganglion neurons in vitro. Sci Rep 7(1):3506. https://doi.org/10.1038/s41598-017-03810-w

117. Treutlein B, Lee QY, Camp JG, Mall M, Koh W, Shariati SA, Sim S, Neff NF, Skotheim JM, Wernig M, Quake SR (2016) Dissecting direct reprogramming from fibroblast to neuron using single-cell RNA-seq. Nature 534(7607):391–395. https://doi.org/10.1038/nature18323

118. Chanda S, Ang CE, Davila J, Pak C, Mall M, Lee QY, Ahlenius H, Jung SW, Sudhof TC, Wernig M (2014) Generation of induced neuronal cells by the single reprogramming factor ASCL1. Stem Cell Rep 3(2):282–296. https://doi.org/10.1016/j.stemcr.2014.05.020

119. Noda T, Meas SJ, Nogami J, Amemiya Y, Uchi R, Ohkawa Y, Nishimura K, Dabdoub A (2018) Direct reprogramming of spiral ganglion non-neuronal cells into neurons: toward ameliorating Sensorineural hearing loss by gene therapy. Front Cell Dev Biol 6:16. https://doi.org/10.3389/fcell.2018.00016

120. Meas SJ, Zhang CL, Dabdoub A (2018) Reprogramming glia into neurons in the peripheral auditory system as a solution for Sensorineural hearing loss: lessons from the central nervous system. Front Mol Neurosci 11:77. https://doi.org/10.3389/fnmol.2018.00077

121. Akil O, Blits B, Lustig LR, Leake PA (2019) Virally mediated overexpression of glial-derived Neurotrophic factor elicits age- and dose-dependent neuronal toxicity and hearing loss. Hum Gene Ther 30(1):88–105. https://doi.org/10.1089/hum.2018.028

122. Lin LF, Doherty DH, Lile JD, Bektesh S, Collins F (1993) GDNF: a glial cell line-derived neurotrophic factor for midbrain dopaminergic neurons. Science 260(5111):1130–1132

123. Buj-Bello A, Buchman VL, Horton A, Rosenthal A, Davies AM (1995) GDNF is an age-specific survival factor for sensory and autonomic neurons. Neuron 15(4):821–828
124. Trupp M, Ryden M, Jornvall H, Funakoshi H, Timmusk T, Arenas E, Ibanez CF (1995) Peripheral expression and biological activities of GDNF, a new neurotrophic factor for avian and mammalian peripheral neurons. J Cell Biol 130(1):137–148. https://doi.org/10.1083/jcb.130.1.137
125. Nosrat CA, Tomac A, Lindqvist E, Lindskog S, Humpel C, Stromberg I, Ebendal T, Hoffer BJ, Olson L (1996) Cellular expression of GDNF mRNA suggests multiple functions inside and outside the nervous system. Cell Tissue Res 286(2):191–207
126. Ylikoski J, Pirvola U, Virkkala J, Suvanto P, Liang XQ, Magal E, Altschuler R, Miller JM, Saarma M (1998) Guinea pig auditory neurons are protected by glial cell line-derived growth factor from degeneration after noise trauma. Hear Res 124(1–2):17–26
127. Yagi M, Kanzaki S, Kawamoto K, Shin B, Shah PP, Magal E, Sheng J, Raphael Y (2000) Spiral ganglion neurons are protected from degeneration by GDNF gene therapy. J Assoc Res Otolaryngol 1(4):315–325. https://doi.org/10.1007/s101620010011
128. Maruyama J, Miller JM, Ulfendahl M (2008) Glial cell line-derived neurotrophic factor and antioxidants preserve the electrical responsiveness of the spiral ganglion neurons after experimentally induced deafness. Neurobiol Dis 29(1):14–21. https://doi.org/10.1016/j.nbd.2007.07.026
129. Glueckert R, Bitsche M, Miller JM, Zhu Y, Prieskorn DM, Altschuler RA, Schrott-Fischer A (2008) Deafferentation-associated changes in afferent and efferent processes in the guinea pig cochlea and afferent regeneration with chronic intrascalar brain-derived neurotrophic factor and acidic fibroblast growth factor. J Comp Neurol 507(4):1602–1621. https://doi.org/10.1002/cne.21619
130. Parker MA (2011) Biotechnology in the treatment of sensorineural hearing loss: foundations and future of hair cell regeneration. J Speech Lang Hear Res 54(6):1709–1731. https://doi.org/10.1044/1092-4388(2011/10-0149)
131. Takahashi K, Yamanaka S (2006) Induction of pluripotent stem cells from mouse embryonic and adult fibroblast cultures by defined factors. Cell 126(4):663–676. https://doi.org/10.1016/j.cell.2006.07.024
132. Johnson KR, Gagnon LH, Tian C, Longo-Guess CM, Low BE, Wiles MV, Kiernan AE (2018) Deletion of a long-range Dlx5 enhancer disrupts inner ear development in mice. Genetics 208(3):1165–1179. https://doi.org/10.1534/genetics.117.300447
133. Li H, Roblin G, Liu H, Heller S (2003) Generation of hair cells by stepwise differentiation of embryonic stem cells. Proc Natl Acad Sci U S A 100(23):13495–13500. https://doi.org/10.1073/pnas.2334503100
134. Oshima K, Shin K, Diensthuber M, Peng AW, Ricci AJ, Heller S (2010) Mechanosensitive hair cell-like cells from embryonic and induced pluripotent stem cells. Cell 141(4):704–716. https://doi.org/10.1016/j.cell.2010.03.035
135. Ouji Y, Sakagami M, Omori H, Higashiyama S, Kawai N, Kitahara T, Wanaka A, Yoshikawa M (2017) Efficient induction of inner ear hair cell-like cells from mouse ES cells using combination of Math1 transfection and conditioned medium from ST2 stromal cells. Stem Cell Res 23:50–56. https://doi.org/10.1016/j.scr.2017.06.013
136. Koehler KR, Mikosz AM, Molosh AI, Patel D, Hashino E (2013) Generation of inner ear sensory epithelia from pluripotent stem cells in 3D culture. Nature 500(7461):217–221. https://doi.org/10.1038/nature12298
137. Costa A, Sanchez-Guardado L, Juniat S, Gale JE, Daudet N, Henrique D (2015) Generation of sensory hair cells by genetic programming with a combination of transcription factors. Development (Cambridge, England) 142(11):1948–1959. https://doi.org/10.1242/dev.119149
138. Schaefer SA, Higashi AY, Loomis B, Schrepfer T, Wan G, Corfas G, Dressler GR, Duncan RK (2018) From otic induction to hair cell production: Pax2(EGFP) cell line illuminates key stages of development in mouse inner ear organoid model. Stem Cells Dev 27(4):237–251. https://doi.org/10.1089/scd.2017.0142

139. Chen W, Jongkamonwiwat N, Abbas L, Eshtan SJ, Johnson SL, Kuhn S, Milo M, Thurlow JK, Andrews PW, Marcotti W, Moore HD, Rivolta MN (2012) Restoration of auditory evoked responses by human ES-cell-derived otic progenitors. Nature 490(7419):278–282. https://doi.org/10.1038/nature11415

140. Ronaghi M, Nasr M, Ealy M, Durruthy-Durruthy R, Waldhaus J, Diaz GH, Joubert LM, Oshima K, Heller S (2014) Inner ear hair cell-like cells from human embryonic stem cells. Stem Cells Dev 23(11):1275–1284. https://doi.org/10.1089/scd.2014.0033

141. Chen JR, Tang ZH, Zheng J, Shi HS, Ding J, Qian XD, Zhang C, Chen JL, Wang CC, Li L, Chen JZ, Yin SK, Shao JZ, Huang TS, Chen P, Guan MX, Wang JF (2016) Effects of genetic correction on the differentiation of hair cell-like cells from iPSCs with MYO15A mutation. Cell Death Differ 23(8):1347–1357. https://doi.org/10.1038/cdd.2016.16

142. Tang ZH, Chen JR, Zheng J, Shi HS, Ding J, Qian XD, Zhang C, Chen JL, Wang CC, Li L, Chen JZ, Yin SK, Huang TS, Chen P, Guan MX, Wang JF (2016) Genetic correction of induced pluripotent stem cells from a deaf patient with MYO7A mutation results in morphologic and functional recovery of the derived hair cell-like cells. Stem Cells Transl Med 5(5):561–571. https://doi.org/10.5966/sctm.2015-0252

143. Koehler KR, Nie J, Longworth-Mills E, Liu XP, Lee J, Holt JR, Hashino E (2017) Generation of inner ear organoids containing functional hair cells from human pluripotent stem cells. Nat Biotechnol 35(6):583–589. https://doi.org/10.1038/nbt.3840

144. Chen J, Hong F, Zhang C, Li L, Wang C, Shi H, Fu Y, Wang J (2018) Differentiation and transplantation of human induced pluripotent stem cell-derived otic epithelial progenitors in mouse cochlea. Stem Cell Res Ther 9(1):230. https://doi.org/10.1186/s13287-018-0967-1

145. Czajkowski A, Mounier A, Delacroix L, Malgrange B (2019) Pluripotent stem cell-derived cochlear cells: a challenge in constant progress. Cell Mol Life Sci 76(4):627–635. https://doi.org/10.1007/s00018-018-2950-5

146. Takeda H, Dondzillo A, Randall JA, Gubbels SP (2018) Challenges in cell-based therapies for the treatment of hearing loss. Trends Neurosci 41(11):823–837. https://doi.org/10.1016/j.tins.2018.06.008

147. Hildebrand MS, Dahl HH, Hardman J, Coleman B, Shepherd RK, de Silva MG (2005) Survival of partially differentiated mouse embryonic stem cells in the scala media of the guinea pig cochlea. J Assoc Res Otolaryngol 6(4):341–354. https://doi.org/10.1007/s10162-005-0012-9

148. Parker MA, Corliss DA, Gray B, Anderson JK, Bobbin RP, Snyder EY, Cotanche DA (2007) Neural stem cells injected into the sound-damaged cochlea migrate throughout the cochlea and express markers of hair cells, supporting cells, and spiral ganglion cells. Hear Res 232(1–2):29–43. https://doi.org/10.1016/j.heares.2007.06.007

Immune System and Macrophage Activation in the Cochlea: Implication for Therapeutic Intervention

Bo hua Hu and Celia Zhang

1 Introduction

Cochlear pathogenesis is a major cause of hearing impairment in the adult population. This degenerative process involves sensory cells, one of the most vulnerable types of cells in the cochlea. Because mammalian cochleae lack the ability to regenerate damaged sensory cells, hair cell death leads to permanent hearing loss. Thus, the prevention of sensory cell death is critical to prevent hearing loss.

The cochlea has intrinsic immune components, and this immune capacity is a major contributor to sensory cell homeostasis and disease. Cochlear immune activation occurs in almost all forms of cochlear disease conditions, including acoustic injury, ototoxicity, sudden deafness, cochlear implantation injury, immune challenges, genetic defects, and age-related degeneration [1–9]. While the precise roles of immune responses in cochlear homeostasis and pathogenesis are poorly understood, both clinical and laboratory studies have revealed the beneficial effects of anti-inflammatory therapeutics in certain cochlear stress conditions [10–18]. In fact, anti-inflammatory treatment is among the few pharmacological options available for ameliorating inner ear disorders. Despite the clinical importance of the cochlear immune system, our knowledge of the fundamental mechanisms governing the cochlear immune reaction to stress is still lacking. This gap is partly due to the lack of biological assays to assess human cochlear tissues. In recent years, animal studies have provided novel insights into the cochlear immune system. In this review, we will provide an overview of these advances with a focus on cochlear immune capacity, immune responses to diseases, and clinical implications for the development of therapeutic interventions.

B. h. Hu (✉) · C. Zhang
Center for Hearing and Deafness, University at Buffalo, Buffalo, NY, USA
e-mail: bhu@buffalo.edu; celiazha@buffalo.edu

© Springer Nature Switzerland AG 2020
S. Pucheu et al. (eds.), *New Therapies to Prevent or Cure Auditory Disorders*,
https://doi.org/10.1007/978-3-030-40413-0_5

2   Overview of the Immune System

The immune system is a host defense system comprising both cellular and molecular components that work together to protect our bodies against infection, disease, and tissue damage. The immune system contains an innate immune system and an adaptive immune system, each with its own cellular and molecular components that participate in specialized tasks (Fig. 1). The innate immune system is the first line of defense, responsible for rapid immune action against nonspecific tissue damage. The main purpose of this response is to restrict the spread of foreign pathogens and eventually remove them. The innate immune system can distinguish foreign molecules from its own based on pathogen-associated molecular patterns (PAMPs) and damage-associated molecular patterns (DAMPs) presented by foreign pathogens and damaged tissue. When hazardous molecules are detected, the innate immune system launches a cascade of immune reactions, including the release of cytotoxic molecules to disrupt pathogens and activate the phagocytic activity of phagocytes to clear pathogens and damaged cells. Once the triggering event is no longer present, the innate immune system facilitates tissue repair. The major function of the innate

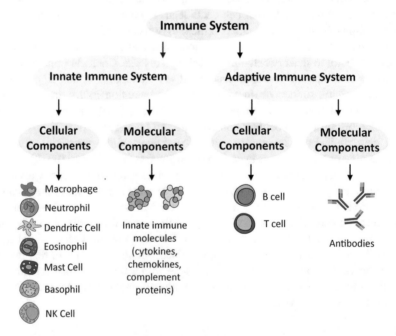

Fig. 1 Schematic illustrating the cellular and molecular components of the innate and adaptive immune system. Cells of the innate immune system include macrophages, neutrophils, dendritic cells, eosinophils, mast cells, basophils, and NK cells. These cells produce immune molecules such as cytokines, chemokines, and complement proteins. Cells of the adaptive immune system include B cells and T cells. These cells produce antibodies that target specific antigens

reaction is mediated by innate immune cells which includes monocytes, macrophages, dendritic cells, neutrophils, basophils, eosinophils, natural killer (NK) cells, and mast cells. The activities of these cells and their trafficking are controlled by immune molecules, such as cytokines and chemokines. These molecules have a variety of immune functions involved in promoting or suppressing inflammation, as well as nonimmune functions. In addition to cytokines and chemokines, the innate response also involves the complement system, which is made of a group of complement proteins. These molecules work with other innate immune mediators to help immune cells clear foreign bodies and promote inflammation through a cascade of complement activities, including opsonizing foreign cells and molecules for detection, attracting macrophages for phagocytosis, lysing cell membrane for pathogen killing, and aggregating pathogens for effective removal. Ideally, the innate immune reaction kills pathogens and confines tissue damage. Once the task is completed, the resolution process proceeds. Complete resolution of inflammation leads to the restoration of tissue homeostasis.

If the innate immune system is unsuccessful in limiting pathogen activities or tissue damage, the adaptive immune response is activated. The adaptive immune response, also known as the acquired immune response, is a much more specific and sophisticated defense response. An advantage of the adaptive immune system is that it has the capacity to remember the molecular signature of the pathogen in order to launch a strong and specific response upon future encounters. Adaptive immune responses are carried out by lymphocytes, including T cells and B cells. Both T cells and B cells are derived from multipotent hematopoietic stem cells in the bone marrow. These immune cells mediate two distinct adaptive immune responses: antibody-mediated and cell-mediated immune responses. In an antibody-mediated response, a naive B cell is activated when it encounters an antigen. The activated B cell proliferates and secretes the antibody that specifically targets the antigen that stimulates its production. Interaction of an antibody and its corresponding antigen blocks the function of the antigen, which is cytotoxic if the antigen is a virus, a microbial toxin, or an aberrant tissue molecule. Antibody binding also marks toxic pathogens for clearance by phagocytic cells of the innate immune system. The cell-mediated immune response involves T cells. T cells are generated in the bone marrow and matured in the thymus gland. Activated T cells react directly against targeted antigens that are present on the surface of antigen-presenting cells. T cells release signal molecules that attract macrophages to engulf the invading microbes or aberrant tissue cells. Overall, the innate and adaptive immune systems work together to take on immune tasks, leading to highly efficient recognition and clearance of pathogens and damaged cells.

Every organ in our body has its unique immune anatomy with specialized immune cells. The inner ear was previously thought to be an "immune-privileged" organ due to the existence of the blood–labyrinth barrier which separates the cochlear microenvironment from the systemic immune system. However, increasing evidence has demonstrated that the cochlea is an immunocompetent organ. Under resting conditions, the cochlea houses endogenous immune components; under acute and chronic stress conditions, the cochlea recruits systemic immune cells. Consequently, the cochlear immune system plays an important role in the pathogenesis of inner ear diseases with various etiologies, serving as a therapeutic target in these disease conditions.

3   Immune Anatomy of the Resting Cochlea

Many organs in the human body have specialized functions that require protection against systemic influence. These organs include the cochlea where the maintenance of a stable environment is essential for hearing function. In the cochlea, the blood–labyrinth barrier prevents the entry of large molecules and cells including those from the systemic immune system. Although the cochlea is isolated from the systemic immune system, it has intrinsic immune components that provide effective immune surveillance under resting conditions enabling it to launch a rapid immune reaction after pathological insults. This intrinsic immune capacity is mediated by cellular and molecular components of the cochlear immune system, both participating in the innate and adaptive immune responses.

3.1   Immune Molecules

The cochlear innate defense relies on a range of innate and adaptive immune molecules, including cytokines, chemokines, pattern recognition molecules, adhesion molecules, and complement molecules. Using RNA-Seq (RNA sequencing), we identified the constitutive expression of more than 1000 genes that have immune functions in the rodent cochlea [19–21]. These genes account for more than 80% of known immune and inflammation-related genes in the body. Furthermore, a large portion of overexpressed genes observed after cochlear damage have immune and inflammatory functions [19–25]. These genes have roles in a variety of immune activities, including inflammation, phagocytosis, complement activation, and antigen presentation. Constitutive expression of a large number of these immune-related genes suggests a critical contribution of the immune system to maintain cochlear integrity.

Immune molecules are expressed by both immune cells and resident tissue cells in the cochlea. In the organ of Corti, supporting cells are the major source of immune molecules [20], suggesting that these cells play a role in immune homeostasis of the sensory cell microenvironment. In the lateral wall, fibrocytes are immune active and express immune mediators, including ICAM-1 [24, 26, 27], an adhesion molecule that facilitates the recruitment of circulating monocytes into the cochlea under stress conditions. Expression levels of these constitutively expressed genes are subject to change when cochlear homeostasis is altered. They can also be modulated by circadian rhythm, a potential risk factor for increased circadian-dependent susceptibility to cochlear stressors [28].

3.2   Cellular Components of the Cochlear Immune System

Outside the cochlea, immune cell composition is known to be organ-dependent with individual organs having their unique immune cell profile. While immune cells have been identified in the human cochlea for a long time, their makeup is still not completely clear. In this regard, studies using mouse cochleae provide us with a useful reference point. Using fluorescence-activated cell sorting, Matern et al. revealed the presence of multiple types of immune cells in postnatal cochleae of Gfil^{Cre} mice [29]. These cells include macrophages (CD11b$^+$, Gr1$^-$), granulocytes (CD11b$^+$, Gr1$^+$), T cells (CD3$^+$), B cells (B220$^+$), and NK cells (CD3$^-$, DX5$^+$), with macrophages accounting for more than 81% of total identified leukocytes. The finding of a macrophage-dominated immune population is consistent with the data derived from immunohistological studies using multiple macrophage markers, including Iba1, CD68, and F4/80 [6, 17, 26, 30, 31]. However, the identification of adaptive immune cells under normal conditions has not been confirmed in immunohistological analyses. For T cell identification, immunohistochemical staining with multiple T cell markers (CD3, CD4, CD8, Lyt-1, and Lyt-2) reveals no positive cells in the cochlea under normal conditions [6, 32, 33]. This discrepancy could be explained in part by differences in T cell markers used, as well as differences in the genetic background and age of the mice used in different studies. It is also possible that the disagreement stems from differences in tissue composition. Cell sorting uses tissues that may contain leukocytes from cochlear blood vessels and the cochlear bony capsule which contains the bone marrow. Immunohistology, on the other hand, largely exclude these cells. In addition to T cells, we have examined mouse cochleae for neutrophils (Ly6G), B cells (CD19, B220), and NK cells (NCR1); no positive cells were found under physiological conditions (our unpublished data). Together, these observations implicate macrophages as the primary effector in the resting cochlea.

The local environment where macrophages reside dictates the functional state of macrophages. In the cochlea, macrophages have been identified in all major anatomic sites, including the lateral wall, spiral limbus, basilar membrane, and neural regions within the modiolus and osseous spiral lamina [34]. Macrophages are also found along the microvessels of the cochlea where they participate in the formation of the blood–labyrinth barrier [35, 36]. In the organ of Corti where sensory cells reside, macrophages are identified in human, but not in mouse cochleae. We are not sure whether this reflects a difference in species or is associated with other factors. One explanation could come from our recent observation of the immune development of postnatal cochleae [34]. In this study, we identified macrophages within the organ of Corti in mouse cochleae during the period of postnatal development. These developmental macrophages die after the cochlea matures, leaving residual cell bodies to be preserved in situ for a long time. If this developmental event occurs in humans, the residual bodies of dead macrophages could be mistakenly identified as intact cells because they maintain immunoreactivity of certain immune cell markers (e.g., CD45).

3.3 Macrophage Origin

Outside the cochlea, tissue macrophages are known to originate from three sources: embryonic origin, local self-renewal, and circulating monocytes [37, 38]. The origin of cochlear macrophages is not completely clear. Immunohistological observations have revealed the presence of cochlear macrophages during the embryonic period in both humans and rodents [39, 40], suggesting that tissue macrophages colonize the cochlea at an early stage of inner ear development. These cells could contribute to the embryonic development of the cochlea. At birth, macrophages have already populated all anatomic compartments where adult macrophages reside [34]. As mentioned above, macrophages are present in the organ of Corti in the postnatal cochleae of mice. These macrophages undergo developmental death via apoptosis as the ear matures. This process starts from the basal end of the cochlea and progresses toward the apex, correlating with the developmental sequence of the organ of Corti. This developmental pattern suggests that macrophages have an important role in the maturation process of the organ of Corti.

3.4 Macrophage Diversity

Macrophages are known to be a heterogeneous group of cells with diverse functional states and phenotypic characteristics [41, 42]. Cochlear macrophages include two dynamically balanced subpopulations: infiltrating and resident. Infiltrating macrophages enter the cochlea after acute damage [6, 8, 13, 17, 24, 26, 43], while resident macrophages reside in the cochlea under steady-state conditions. These cells are phenotypically and functionally plastic in response to changes in their microenvironment. Within the cochlea, different partitions have different cellular compositions that create different microenvironments for macrophages. Therefore, macrophages at different anatomic sites are expected to have site-specific functions in homeostatic conditions with unique response patterns under stress. Indeed, cochlear macrophages at different partitions display distinct location-specific phenotypes and expression patterns of immune molecules. In addition, macrophages that reside in the same anatomic site can also display phenotypic differences. For instance, macrophages on the luminal surface of the scala tympani display diverse morphologies, from rounded to branched to dendritic shape and from a small to a large size. As demonstrated in Fig. 2, rounded cells are more phagocytic. Even cells with the same morphology in the same location can display distinct expression pattern of immune proteins. For example, a small group of immune cells in the apical portion of the cochlea express major histocompatibility complex class II (MHC-II) [33], an antigen-presenting protein that transfers peptides to T cell receptors for the activation of the adaptive immune system. These positive cells have a dendritic shape and intermingle with MHC-II negative cells, which have the same morphology. The finding of MHC-II expression in a subgroup of macrophages in the same

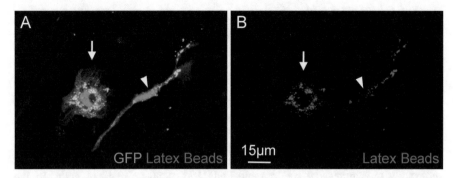

Fig. 2 Macrophage phagocytosis in the cochlea. Fluoresbrite® Polychromatic latex beads were used to determine the level of macrophage phagocytic activity with different morphologies. (**A–B**) The arrow points to a GFP-labeled macrophage (green) with an amoeboid shape that contains a large number of latex beads (red). In contrast, the macrophage with a long spindle shape (pointed by the arrowhead) contains only a few latex beads. The cochlea was collected from a CX3CR1-GFP mouse at 1 day after exposure to an intense noise of 120 dB SPL for 1 h

location with the same morphology suggests the existence of macrophage subsets that have differential physiological activities.

On the surface of the basilar membrane, the morphology of macrophages corresponds with their location along the cochlear spiral [33, 44]. In the apical section, macrophages are dendritic with thin and long processes. In the basal and middle sections, however, macrophages display a branched or an amoeboid shape with several short processes or no projections at all. This apex-to-base transition is accompanied by site-specific variations in immune molecule expression. Under resting conditions, CD11c and CD14, both having functions in inflammatory activities, are expressed in the basal macrophages [33]. This phenotypic profile suggests that basal macrophages are pro-inflammatory, which could render them susceptible to inflammatory activation under stress. This intrinsic pro-inflammatory state in the basal end of the cochlea could serve as an underlying mechanism for the greater susceptibility of basal sensory cells in aging degeneration, acoustic trauma, ototoxicity, and many other pathological conditions.

4   Cochlear Inflammation

Inflammation is a host defense mechanism against extrinsic pathogens and intrinsic tissue damage. Based on underlying causes, tissue inflammation can be broadly classified into infection inflammation and sterile inflammation. Primary infection is rare in the cochlea due to its protected location and the blood–labyrinth barrier. However, secondary inflammation could occur as a result of an infection in neighboring tissues when the infection spreads into the cochlea via the direct dissemination of inflammatory molecules and bacteria [45–47]. Meningitis is initiated in the central nervous system. However, peripheral spreading via the cochlear aqueduct causes acute inflammation in the cochlea and sensorineural hearing loss [48, 49]. Otitis media can spread to the inner ear via the round window, causing sensorineural hearing loss. Nevertheless, sterile inflammation is more common in the cochlea. Early studies of sterile inflammation focused on instances caused by immune-mediated reactions [50–52]. Recent studies have shifted the attention to a disease-induced inflammatory reaction, including acoustic trauma, cochlear implantation injury, and ototoxicity.

Acute inflammation has two distinct but inseparable stages: induction and resolution (Fig. 3). This inflammatory cascade is mediated by immune cells, and immune molecules biosynthesized by immune cells and cochlear resident cells. In the cochlea, sensory cells are the most vulnerable cell type and, thus, sensory cell damage is a common trigger of cochlear inflammation. Mounting evidence has shown that stress signals such as free radicals and cytokines activate the inflammatory response via the nuclear factor kappa-light-chain enhancer of activated B cells (NF-κB) signaling pathway [53, 54]. NF-κB is a family of transcription factors that regulate the transcription of a set of immune-related genes including chemokines, cytokines, pro-inflammatory enzymes, and adhesion molecules. Sensory cell damage can also cause the release of tissue injury-induced ligands that act as DAMP

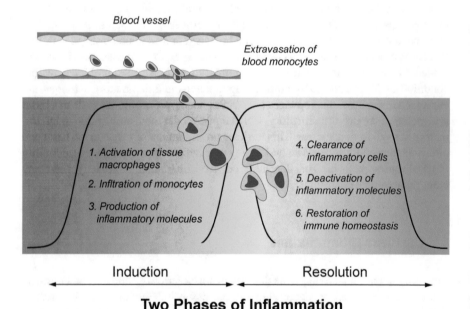

Two Phases of Inflammation

Fig. 3 Schematic illustrating the two phases of inflammation: induction and resolution. During the induction phase, tissue macrophages are activated, and blood monocytes infiltrate the tissue, both producing inflammatory molecules. Once the triggering event is cleared, inflammation begins to resolve. The resolution process involves the clearance of inflammatory cells and deactivation of inflammatory molecules, which eventually leads to the restoration of tissue homeostasis

molecules to interact with immune-sensing receptors. The toll-like receptors (TLRs) are a family of damage-sensing receptors that are important components of the innate immune system. These receptors are involved in the detection of both exogenous molecules from pathogens and endogenous molecules from damaged tissues, including lipopolysaccharides from the outer membrane of bacteria and heat shock proteins, components of the extracellular matrix, fibrinogen, high mobility group box 1 (HMGB-1), β-defensin, and surfactant protein A (SP-A) [55–58]. Upon binding to their ligands, TLRs recruit adaptor molecules and kinases to activate all aspects of the immune and inflammatory responses, including the production of inflammatory molecules and activation of immune cells via the NF-κB signaling pathway [59–61]. In the cochlea, our comprehensive screening of TLR family members has unveiled the constitutive expression of TLR2, TLR3, and TLR4 [19–21, 62]. Among the expressed TLRs, TLR3 and TLR4 are inducible after acoustic injury. Importantly, the expression of TLRs has been found in supporting cells adjacent to damaged sensory cells after acoustic injury [20]. While the detailed downstream events after activation of the TLR signaling pathway are still unclear, we have demonstrated that *Tlr4* knockout alters the inflammatory response to acoustic trauma and reduces the level of noise-induced sensory cell damage. Conversely, studies from other laboratories have revealed that activation of the TLR4 pathway by the treatment of lipopolysaccharide, a ligand of TLR4, potentiates the level of sensory cell damage induced by ototoxic drugs [2, 63]. Because TLR4 signaling can be modulated by several small molecules [64], therapeutic modulation of TLR function is possible and may hold promise for the prevention of inflammatory damage.

4.1   Production of Inflammatory Molecules

RNA-Seq provides precise measurements of the entire transcriptome in a high-throughput and quantitative fashion. We have used this technique to survey the expression changes of cochlear genes after acoustic injury [19, 21]. This analysis identifies a group of differentially expressed genes that function in signaling pathways linking immune activities such as the chemokine signaling pathway, the cytokine-cytokine receptor pathway, the Toll-receptor signaling pathway, and the complement pathway [21]. Together with the data collected using other expression analyses, we now start to understand the involvement of common pro-inflammatory molecules. Tumor necrotic factor (TNF) is a multifunctional cytokine that plays important roles in diverse cellular events, including inflammation. This molecule is produced by macrophages, monocytes, and neutrophils upon stimulation. In the cochlea, its expression is rapidly induced after acoustic injury [31] and immune challenge [23]. TNF has been shown to provoke infiltration of inflammatory cells into the cochlea [65]. Consistent with this finding, the inhibition of TNF has been found to be effective for ameliorating the cochlear inflammation caused by immune challenge with the intracochlear administration of keyhole limpet hemocyanin [23, 66]. In addition, interleukin-6 (IL-6) and interleukin-1β (IL-1β) are expressed at low levels in the cochlea under resting conditions, and their expression levels are upregulated in response to acoustic trauma [17, 31, 67] and immune-mediated challenge [23]. Both IL-6 and IL-1β are potent pro-inflammatory cytokines that are produced by immune cells upon activation. Again, suppression of these pro-inflammatory cytokines with an anti-IL-6 antibody or a heat shock protein inducer (geranylgeranylacetone) has been found to be effective for reducing cochlear inflammation and sensory cell damage [17, 67]. The expression of chemokine (C–C motif) ligand 2 (CCL2), also known as monocyte chemoattractant protein 1 (MCP-1) [68], has been found after acoustic injury in cochlear tissues [19, 26] and activated macrophages [69]. However, *Ccl2* knockout does not affect monocyte migration, nor the level of sensory cell pathogenesis after acoustic overstimulation [13]. This result contradicts the finding that although *Ccr2* knockout does not alter the level of monocyte infiltration [13], disruption of the CCL2 receptor potentiates the magnitude of sensory cell pathogenesis induced by acoustic injury.

The inflammatory cascade is a dynamic process with clear phasic characteristics after acute cochlear stress. Animal studies have shown that the induction of acute inflammation is a rapid process characterized by the production of pro-inflammatory mediators that peaks 1 day after noise exposure [31, 62]. However, this early wave of inflammatory molecule production quickly subsides even though sensory cell lesions continue to grow. Interestingly, a second peak of inflammatory molecule production has been found several days after noise exposure [27]. While the biological basis for this second wave of inflammation is unclear, the continuous production of pro-inflammatory molecules could lead to prolonged inflammation.

4.2   Recruitment of Macrophages

In addition to inflammatory molecules, acute cochlear inflammation involves inflammatory cells including both resident and recruited macrophages. As discussed above, resident macrophages reside in the cochlea and play a surveillance role under resting conditions. Acute cochlear stress activates these inflammatory sensors, leading to the expression of inflammatory molecules including CCL2, a monocyte chemoattractant molecule. These chemoattractants cause an expansion of the immune cell population. Early studies identify these cells as macrophages based on their morphological features [43, 70]. Their identity was subsequently confirmed with macrophage-specific markers [6, 52].

The expansion of the macrophage population has been found in virtually all types of acute cochlear stress conditions. This expansion is achieved primarily by recruiting circulating monocytes through the cochlear vasculature. Circulating monocytes comprise three subsets: classical, intermediate, and nonclassical, each with distinct inflammatory activities [71, 72]. In humans, classical monocytes are CD14++/CD16−, intermediate monocytes are CD14++/CD16+, and nonclassical monocytes are CD14+/CD16++. In mice, classical monocytes are Ly6C^{high}, whereas nonclassical monocytes are Ly6C^{low}. When tissue inflammation occurs, Ly6C^{high}

cells enter the site of inflammation and participate in the inflammatory reaction. In contrast, Ly6C^{low} monocytes adhere to and move along the endothelium of blood vessels, clearing damaged cells and triggering inflammatory responses without entering the tissue [73, 74]. In noise-damaged cochleae, we found that cells with monocyte-like phenotypes display strong Ly6C immunoreactivity [69]. This finding suggests that pro-inflammatory monocytes enter the cochlea during acute inflammation. Once in the tissue, monocytes differentiate into activated macrophages and produce immune-related molecules including CCL2 and MHC-II [33, 69].

Similar to the production of inflammatory molecules, macrophage activity displays distinct phasic characteristics. In animal models of acoustic trauma, macrophage infiltration has been found to start at 1 day and peak at 3–5 days after injury [6, 17, 26, 43]. This onset of macrophage recruitment apparently lags behind the early production of inflammatory molecules, which occurs within 1 day after noise injury [31, 62]. This interval could represent the time required for the interaction between resident macrophages that release chemokines and infiltrating cells that are attracted to the chemokines. The arrival of macrophages occurs when sensory cell death has already taken place, which creates the need for macrophages to clear damaged cells, thus facilitating tissue repair.

4.3   Clearance of Inflammatory Cells

Acute cochlear inflammation is a self-limiting process. Once the triggering event has passed, inflammation begins to resolve. The cellular mechanism for deactivating inflammatory cells in the cochlea is unclear. Three exit routes available to inflammatory cells have been identified in non-cochlear tissues: (1) in situ death [75, 76], (2) systemic recirculation (inflammatory cells return to the bloodstream) [77], and (3) lymphatic drainage [78]. Although lymphatic drainage of trace molecules from the cochlea has been reported [79], the presence of a lymphatic system has not been characterized. We recently reexamined this issue using an antibody against Lyve-1, a lymphatic endothelial cell marker that has been successfully used for identifying lymphatic vessels in the brain [80]. However, we did not find any Lyve-positive vessels in the cochlea (unpublished observation). For immune cell recirculation, no evidence has been reported for the existence of this exit route. Therefore, it is likely that local cell death is the major route of inflammatory cell clearance. Local cell death could provoke additional inflammatory activities and require an additional phagocytic activity for clearance. As a result, the intracochlear deactivation of inflammatory cells could provoke a vicious cycle of inflammation, delaying the recovery process. A better understanding of the removal process of inflammatory cells is needed for promoting a silent clearance of unwanted immune cells and thus a speedy resolution of inflammation.

4.4   Chronic Disease, Aging, and Inflammation

Chronic inflammation stems from persistent pathogenic stimuli. Chronic cochlear pathogeneses such as age-related cochlear pathologies and genetic defects lead to a slow progressive sensory cell death. In these disease conditions, sensory cell lesions usually start from the basal end of the cochlea and then progresses toward the apical region. Unlike acute inflammation that has a distinct onset and offset, chronic inflammation is typically persistent with no clear onset or offset. We have found that macrophages acquire activated phenotypes before the onset of sensory cell death during age-related degeneration [44]. In this situation, macrophages in the basal end of the cochlea transform from a branched to an amoeboid morphology, a phenotypic characteristic of pro-inflammation activation [81–83]. With the onset and progression of sensory cell pathogenesis, this activated morphology spreads toward the apex, preceding sensory cell degeneration. The fact that the cochlear immune environment precedes sensory cell death emphasizes the potential contribution of the immune environment to sensory cell homeostasis. Therefore, modulating the cochlear immune state may have therapeutic potential for reducing the progression of sensory cell pathogenesis in chronic disease conditions.

5   Potential Roles of the Immune System in Cochlear Homeostasis and Disease

Accumulated evidence has revealed that immune integrity is important for cochlear homeostasis. Aberrant immune activity could contribute to the development of cochlear diseases. Recent studies have revealed the functional impacts of genetic defects in immune-related genes on the auditory system. For example, the lack of expression of the major histocompatibility complex class I genes, *H2-Kb* and *H2-Db,* has been found to cause high-frequency hearing impairment [84]. However, in many conditions, deficiencies in immune-related genes do not affect hearing sensitivity under resting condition but alter the cochlea's susceptibility to stress. *Ccr2* knockout has been reported to potentiate noise-induced cochlear damage [13]. In contrast, *Tlr4* knockout increases the tolerance of sensory cells to acoustic overstimulation [62]. These opposing effects of immune disruption highlight the diverse roles of immune-related genes in the regulation of cochlear responses to stress.

5.1   Immune Surveillance

Immune surveillance is an essential function of the immune system for the maintenance of tissue homeostasis. This function relies on resident macrophages. As an immune sensor, cochlear macrophages reside in all cochlear compartments, and their processes extend over large areas. Based on their location, cochlear macrophages can be classified as embedded macrophages (cell bodies that lie within the tissue) and surface macrophages (cell bodies that lie on the surface of the tissue). Surface macrophages, including those on the luminal surface of the scala tympani, are bathed in the perilymph. This environment imposes fewer physical constraints permitting greater mobility and morphological transformation. Thus, these cells can easily acquire an activated morphology in response to sensory cell pathogenesis. We have observed morphological changes of basilar membrane macrophages under resting conditions as well as during pathogenesis, and we have found that these cells are a sensitive indicator of the cochlear immune state [9, 44, 69].

5.2   Clearance of Damaged Cells

Phagocytosis is an intrinsic function of macrophages. These cells constantly survey their surroundings, looking for external pathogens or endogenous damaged cells. In the cochlea, the entry of pathogens is rare due to the blood–labyrinth barrier. Therefore, the major targets of macrophage engulfment are the damaged cochlear cells and their components. As mentioned earlier, one of the most vulnerable cells in the cochlea are sensory cells in the organ of Corti. However, the organ of Corti lacks resident macrophages. Therefore, macrophages have to migrate to or extend their processes to the region of sensory cell damage. Several studies have demonstrated the entry of macrophages to the organ of Corti after acute damage [40, 43, 85, 86]. However, this migration appears only when extreme damage occurs. Often, such damage results in the disruption of the entire structure of the organ of Corti and the loss of a large number of sensory cells and supporting cells. In such events, macrophages, as the "big eaters" of the immune system, take on the sole responsibility of phagocytosis to remove the cell debris. In the case of chronic sensory cell degeneration, the evidence for migration of macrophages into the site of sensory cell death is still lacking. Instead, supporting cells around the dead sensory cells appear to aid in the removal of the dead cells. This notion is demonstrated by Abrashkin and colleagues who used prestin as a marker to track the fate of outer hair cell debris. Their study revealed the presence of prestin immunoreactivity within Deiters cells after acoustic and ototoxic damage [87], clearly implicating supporting cells in the phagocytosis of sensory cell debris. At present, the molecular mechanism that controls the detection of dead sensory cells by supporting cells remains elusive. We found an increased expression of TLR4 in Deiters cells beneath dying sensory cells [62], suggesting that the TLR signaling pathway is involved in the supporting cell's detection of dead sensory cells. Importantly, we found that *Tlr4* knockout reduced the level of sensory cell damage, suggesting that TLR4-mediated inflammatory activity is detrimental to sensory cell integrity.

5.3   Neuroinflammation

Alongside the sensory cell environment, the cochlear immune system contributes to the neural environment of the cochlea. Cochlear inflammation and its associated macrophage activation have been linked to neuroinflammation in the ganglion cells and their peripheral fibers that innervate sensory cells. Under physiological conditions, macrophages are present in the neuronal region within the osseous spiral lamina and the modiolus [30, 34, 88, 89]. This immune cell pool expands after cochlear damage [13, 26]. Using an animal model of cochlear implantation trauma, Bas et al. reported that mechanical stress caused an increase in macrophage population in the neural region of the cochlea [90]. Noticeably, macrophages acquired activated phenotypes as evidenced by the enhanced expression of a pro-inflammatory molecule, IL-1β. Interestingly, the expression of anti-inflammatory molecules was also observed in activated macrophages, suggesting a complex role of macrophages in the modulation of inflammatory responses to cochlear damage. The precise roles of macrophage activity in neuronal function remain elusive. A recent study has shown that enhanced macrophage activity is protective against neuronal degeneration [86]. In this study, the authors investigated how macrophage dysfunction affected neural survival using the $Cx3cr1^{GFP/GFP}$ mouse. This strain of mice lacks functional fractalkine receptors in macrophages, leading to a disruption of fractalkine signaling that is known to regulate the migration of circulating monocytes [91, 92]. The study found that the loss of fractalkine signaling reduced macrophage response to sensory cell pathogenesis as evidenced by a reduction in macrophage infiltration into the sensory epithelium and the spiral ganglion region. Moreover, $Cx3cr1$-deficient mice display an exacerbated degeneration of spiral ganglion neurons. This observation underscores the importance of macrophages in neuronal responses to sensory cell pathogenesis.

5.4   Tinnitus and Hyperacusis

Patients with cochlear damage often suffer from tinnitus and hyperacusis. Recent studies have focused on the pathophysiology of these debilitating symptoms. While the direct link between these clinical symptoms and neuroinflammation has yet to be established, findings from studies on the role of neuroinflammation in pain raise the possibility that cochlear inflammation could also serve as an underlying mechanism for tinnitus and hyperacusis. This notion is supported by clinical observations that tinnitus often occurs in patients with a history of pathological insults that can cause cochlear inflammation. Moreover, anti-inflammatory treatments are sometimes effective in alleviating tinnitus and hyperacusis in certain patients. In both the peripheral and central nervous systems, it has been found that pro-inflammatory mediators are directly involved in the development of neuropathic pain [93]. Pro-inflammatory cytokines can cause neuronal activation by regulating their activity

[94, 95]. In the cochlea, sensory cell damage causes inflammatory activation not only in the sensory epithelium but also in the neuronal regions [30, 88, 89]. Such inflammatory activity could modulate the activity of spiral ganglion neurons, such as unmyelinated type II afferent neurons that have been linked to the sensation of painfully loud sounds [96].

5.5 Immune Priming

Disruption of immune homeostasis in an organ can alter its response to subsequent stress, a phenomenon known as immune priming [97]. Systemic activation of the innate immune system by inflammatory stimuli, such as lipopolysaccharide and keyhole limpet hemocyanin, sensitizes the cochlea to subsequent pathogenic stress [2, 24, 63]. Specifically, such treatments lead to an augmented cochlear immune reaction and exacerbated pathophysiology of the cochlea. The augmented response is likely mediated by an adaptive immune response, which is activated by the initial challenge [98]. Presently, it is not completely clear whether cochlear damage could serve as a triggering event for altering immune homeostasis and thereby its response to subsequent stress. Prior damage to the cochlea is known to affect cochlear responses to subsequent stress. This interaction depends on the types and levels of the initial and secondary stressors. Several laboratories have demonstrated that pre-treatment with low-level noise renders the cochlea resistant to subsequent acoustic injury and ototoxicity [99–106]. In contrast, a prior history of high-level stress sensitizes the cochlea to future damage [107]. We recently found that unresolved inflammation potentiates the cochlear inflammatory reaction to subsequent acoustic overstimulation [108]. Therefore, preparing the cochlear immune system for stress (priming) could offer therapeutic benefits and warrants further investigation.

6 Immune Modulation for Therapeutic Purpose

Various immune treatments have been developed to control inflammatory reactions and inflammatory damage. Some of the treatments have been found to be effective in treating certain inner ear disorders with immune components [10–18]. Current treatments assume that inflammation is detrimental as it compromises the sensory cell environment and thus aggravates the pathologic outcome. Glucocorticoids are a class of steroid hormones produced in the adrenal cortex. These small lipophilic compounds have strong immunosuppressive and anti-inflammatory effects, as well as other biological effects. Their action requires binding to the endogenous glucocorticoid receptor, a transcription factor that belongs to the nuclear receptor super-family. Glucocorticoid receptors have been found in various cochlear partitions, including the spiral ganglion neurons and the spiral ligament [109, 110]. Activation of these receptors alters the transcription of immune-related genes, thus modulating inflammatory activities in the cochlea. Indeed, treatment with a steroid agent has been found to induce a complex change in the expression of inflammatory genes including downregulation of pro-inflammatory genes, such as IL-1β [111]. Commonly used steroid treatments include synthetic glucocorticoids such as pred- nisone, dexamethasone, and methylprednisolone. These drugs can be applied locally or systemically. Local application yields a high intra-cochlear concentration [112, 113], which not only enhances the effects but also reduces systemic absorption and systemic side effects. In addition to steroidal treatment, nonsteroidal anti- inflammatory drugs have also been tested. These drugs often target a specific inflam- matory event or immune mediator, providing an opportunity for selective intervention. Etanercept is an FDA-approved TNF inhibitor for clinical use in cer- tain inflammatory disorders such as rheumatoid arthritis, axial spondyloarthritis, and psoriatic arthritis [114]. In an animal model of immune-mediated labyrinthitis induced by keyhole limpet hemocyanin, Etanercept has been shown to reduce cochlear inflammatory activities, including the infiltration of inflammatory cells [23, 66]. Moreover, this treatment reduced hearing impairments caused by the inflammation. In clinical studies, targeting TNF with Etanercept and methotrexate was shown to be efficacious in selective patients with immune-mediated inner ear disorders [115, 116]. However, the outcome of the treatment varies significantly among individual patients [117]. This variation suggests the complexity of the dis- ease condition, emphasizing the need for further studies on inflammatory activities underlying clinical symptoms.

7   Future Directions

Presently, modulating inflammation in the cochlea is focused on suppressing inflam- mation. While this strategy is beneficial in alleviating inflammatory activities in certain disease conditions, improvements in the final outcome of the inflammatory process are generally limited. This limitation stems from a lack of understanding of the inflammatory processes and the differential effects of inflammation on cochlear pathogenesis and its repair process. Inflammation is a dynamic process with diverse roles including an essential function for clearing damaged cells and promoting tis- sue repair. Therefore, suppressing this process could minimize these beneficial effects. Recent studies in non-cochlear tissues have demonstrated the benefits of promoting inflammation resolution. Although this strategy has not been thoroughly studied in the cochlea, we believe that it holds great promise for preventing the progression of acute inflammation to chronic inflammation. It is expected that com- bining pro-resolution with anti-inflammatory technology would offer an enhanced, integrated strategy compared with existing approaches. In this regard, the targeted modulation of inflammatory activity is a new research direction for discovering effective treatments for inner ear disorders.

Acknowledgments This work was supported by the National Institute on Deafness and Other Communication Disorders of the National Institutes of Health [R01DC010154 (BHH)].

References

1. Verschuur C, Causon A, Green K, Bruce I, Agyemang-Prempeh A, Newman T (2015) The role of the immune system in hearing preservation after cochlear implantation. Cochlear Implants Int 16(Suppl 1):S40–SS2
2. Hirose K, Li SZ, Ohlemiller KK, Ransohoff RM (2014) Systemic lipopolysaccharide induces cochlear inflammation and exacerbates the synergistic ototoxicity of kanamycin and furosemide. J Assoc Res Otolaryngol 15(4):555–570
3. Goodall AF, Siddiq MA (2015) Current understanding of the pathogenesis of autoimmune inner ear disease: a review. Clin Otolaryngol 40(5):412–419
4. Iwai H, Lee S, Inaba M, Sugiura K, Baba S, Tomoda K et al (2003) Correlation between accelerated presbycusis and decreased immune functions. Exp Gerontol 38(3):319–325
5. Toubi E, Ben-David J, Kessel A, Halas K, Sabo E, Luntz M (2004) Immune-mediated disorders associated with idiopathic sudden sensorineural hearing loss. Ann Otol Rhinol Laryngol 113(6):445–449
6. Hirose K, Discolo CM, Keasler JR, Ransohoff R (2005) Mononuclear phagocytes migrate into the murine cochlea after acoustic trauma. J Comp Neurol 489(2):180–194
7. Gazquez I, Soto-Varela A, Aran I, Santos S, Batuecas A, Trinidad G et al (2011) High prevalence of systemic autoimmune diseases in patients with Meniere's disease. PLoS One 6(10):e26759
8. Warchol ME, Schwendener RA, Hirose K (2012) Depletion of resident macrophages does not alter sensory regeneration in the avian cochlea. PLoS One 7(12):e51574
9. Zhang C, Sun W, Li J, Xiong B, Frye MD, Ding D et al (2017) Loss of sestrin 2 potentiates the early onset of age-related sensory cell degeneration in the cochlea. Neuroscience 361:179–191
10. Psillas G, Pavlidis P, Karvelis I, Kekes G, Vital V, Constantinidis J (2008) Potential efficacy of early treatment of acute acoustic trauma with steroids and piracetam after gunshot noise. Eur Arch Otorhinolaryngol 265(12):1465–1469
11. Zhou Y, Zheng G, Zheng H, Zhou R, Zhu X, Zhang Q (2013) Primary observation of early transtympanic steroid injection in patients with delayed treatment of noise-induced hearing loss. Audiol Neurootol 18(2):89–94
12. Takahashi K, Kusakari J, Kimura S, Wada T, Hara A (1996) The effect of methylprednisolone on acoustic trauma. Acta Otolaryngol 116(2):209–212
13. Sautter NB, Shick EH, Ransohoff RM, Charo IF, Hirose K (2006) CC chemokine receptor 2 is protective against noise-induced hair cell death: studies in CX3CR1(+/GFP) mice. J Assoc Res Otolaryngol 7(4):361–372
14. Canlon B, Meltser I, Johansson P, Tahera Y (2007) Glucocorticoid receptors modulate auditory sensitivity to acoustic trauma. Hear Res 226(1–2):61–69
15. Fakhry N, Rostain JC, Cazals Y (2007) Hyperbaric oxygenation with corticoid in experimental acoustic trauma. Hear Res 230(1–2):88–92
16. Hoshino T, Tabuchi K, Hirose Y, Uemaetomari I, Murashita H, Tobita T et al (2008) The non-steroidal anti-inflammatory drugs protect mouse cochlea against acoustic injury. Tohoku J Exp Med 216(1):53–59
17. Wakabayashi K, Fujioka M, Kanzaki S, Okano HJ, Shibata S, Yamashita D et al (2010) Blockade of interleukin-6 signaling suppressed cochlear inflammatory response and improved hearing impairment in noise-damaged mice cochlea. Neurosci Res 66(4):345–352

18. Takemura K, Komeda M, Yagi M, Himeno C, Izumikawa M, Doi T et al (2004) Direct inner ear infusion of dexamethasone attenuates noise-induced trauma in guinea pig. Hear Res 196(1–2):58–68

19. Yang SZ, Cai QF, Vethanayagam RR, Wang JM, Yang WP, Hu BH (2016) Immune defense is the primary function associated with the differentially expressed genes in the cochlea following acoustic trauma. Hear Res 333:283–294

20. Cai Q, Vethanayagam RR, Yang S, Bard J, Jamison J, Cartwright D et al (2014) Molecular profile of cochlear immunity in the resident cells of the organ of Corti. J Neuroinflammation 11(1):173

21. Patel M, Hu Z, Bard J, Jamison J, Cai Q, Hu BH (2013) Transcriptome characterization by RNA-Seq reveals the involvement of the complement components in noise-traumatized rat cochleae. Neuroscience 248C:1–16

22. Cho Y, Gong TW, Kanicki A, Altschuler RA, Lomax MI (2004) Noise overstimulation induces immediate early genes in the rat cochlea. Brain Res Mol Brain Res 130(1–2):134–148

23. Satoh H, Firestein GS, Billings PB, Harris JP, Keithley EM (2002) Tumor necrosis factor-alpha, an initiator, and etanercept, an inhibitor of cochlear inflammation. Laryngoscope 112(9):1627–1634

24. Miyao M, Firestein GS, Keithley EM (2008) Acoustic trauma augments the cochlear immune response to antigen. Laryngoscope 118(10):1801–1808

25. Han Y, Hong L, Zhong C, Chen Y, Wang Y, Mao X et al (2012) Identification of new altered genes in rat cochleae with noise-induced hearing loss. Gene 499(2):318–322

26. Tornabene SV, Sato K, Pham L, Billings P, Keithley EM (2006) Immune cell recruitment following acoustic trauma. Hear Res 222(1–2):115–124

27. Tan WJ, Thorne PR, Vlajkovic SM (2016) Characterisation of cochlear inflammation in mice following acute and chronic noise exposure. Histochem Cell Biol 146(2):219–230

28. Sarlus H, Fontana JM, Tserga E, Meltser I, Cederroth CR, Canlon B (2019) Circadian integration of inflammation and glucocorticoid actions: implications for the cochlea. Hear Res 377:53–60

29. Matern M, Vijayakumar S, Margulies Z, Milon B, Song Y, Elkon R et al (2017) Gfi1Cre mice have early onset progressive hearing loss and induce recombination in numerous inner ear non-hair cells. Sci Rep 7:42079

30. Okano T, Nakagawa T, Kita T, Kada S, Yoshimoto M, Nakahata T et al (2008) Bone marrow-derived cells expressing Iba1 are constitutively present as resident tissue macrophages in the mouse cochlea. J Neurosci Res 86(8):1758–1767

31. Fujioka M, Kanzaki S, Okano HJ, Masuda M, Ogawa K, Okano H (2006) Proinflammatory cytokines expression in noise-induced damaged cochlea. J Neurosci Res 83(4):575–583

32. Takahashi M, Harris JP (1988) Anatomic distribution and localization of immunocompetent cells in normal mouse endolymphatic sac. Acta Otolaryngol 106(5–6):409–416

33. Yang W, Vethanayagam RR, Dong Y, Cai Q, Hu BH (2015) Activation of the antigen presentation function of mononuclear phagocyte populations associated with the basilar membrane of the cochlea after acoustic overstimulation. Neuroscience 303:1–15

34. Dong Y, Zhang C, Frye M, Yang W, Ding D, Sharma A et al (2018) Differential fates of tissue macrophages in the cochlea during postnatal development. Hear Res 365:110–126

35. Shi X (2010) Resident macrophages in the cochlear blood-labyrinth barrier and their renewal via migration of bone-marrow-derived cells. Cell Tissue Res 342(1):21–30

36. Hirose K, Li SZ (2019) The role of monocytes and macrophages in the dynamic permeability of the blood-perilymph barrier. Hear Res 374:49–57

37. Prinz M, Priller J (2014) Microglia and brain macrophages in the molecular age: from origin to neuropsychiatric disease. Nat Rev Neurosci 15(5):300–312

38. Haldar M, Murphy KM (2014) Origin, development, and homeostasis of tissue-resident macrophages. Immunol Rev 262(1):25–35

39. Kim JH, Rodriguez-Vazquez JF, Verdugo-Lopez S, Cho KH, Murakami G, Cho BH (2011) Early fetal development of the human cochlea. Anat Rec (Hoboken) 294(6):996–1002

40. Hirose K, Rutherford MA, Warchol ME (2017) Two cell populations participate in clearance of damaged hair cells from the sensory epithelia of the inner ear. Hear Res 352:70–81

41. Geissmann F, Gordon S, Hume DA, Mowat AM, Randolph GJ (2010) Unravelling mono-nuclear phagocyte heterogeneity. Nat Rev Immunol 10(6):453–460

42. Gordon S, Taylor PR (2005) Monocyte and macrophage heterogeneity. Nat Rev Immunol 5(12):953–964

43. Fredelius L, Rask-Andersen H (1990) The role of macrophages in the disposal of degen-eration products within the organ of Corti after acoustic overstimulation. Acta Otolaryngol 109(1–2):76–82

44. Frye MD, Yang W, Zhang C, Xiong B, Hu BH (2017) Dynamic activation of basilar mem-brane macrophages in response to chronic sensory cell degeneration in aging mouse cochleae. Hear Res 344:125–134

45. Morizono T, Giebink GS, Paparella MM, Sikora MA, Shea D (1985) Sensorineural hear-ing loss in experimental purulent otitis media due to Streptococcus pneumoniae. Arch Otolaryngol 111(12):794–798

46. Ghaheri BA, Kempton JB, Pillers DAM, Trune DR (2007) Cochlear cytokine gene expres-sion in murine acute otitis media. Laryngoscope 117(1):22–29

47. Ichimiya I, Suzuki M, Hirano T, Mogi G (1999) The influence of pneumococcal otitis media on the cochlear lateral wall. Hear Res 131(1–2):128–134

48. Kesser BW, Hashisaki GT, Spindel JH, Ruth RA, Scheld WM (1999) Time course of hear-ing loss in an animal model of pneumococcal meningitis. Otolaryngol Head Neck Surg 120(5):628–637

49. Klein M, Koedel U, Pfister HW, Kastenbauer S (2003) Morphological correlates of acute and permanent hearing loss during experimental pneumococcal meningitis. Brain Pathol 13(2):123–132

50. Woolf NK, Harris JP (1986) Cochlear pathophysiology associated with inner ear immune responses. Acta Otolaryngol 102(5–6):353–364

51. Ma C, Billings P, Harris JP, Keithley EM (2000) Characterization of an experimentally induced inner ear immune response. Laryngoscope 110(3 Pt 1):451–456

52. Takahashi M, Harris JP (1988) Analysis of immunocompetent cells following inner ear immunostimulation. Laryngoscope 98(10):1133–1138

53. So H, Kim H, Lee JH, Park C, Kim Y, Kim E et al (2007) Cisplatin cytotoxicity of audi-tory cells requires secretions of proinflammatory cytokines via activation of ERK and NF-kappaB. J Assoc Res Otolaryngol 8(3):338–355

54. Adams JC, Seed B, Lu N, Landry A, Xavier RJ (2009) Selective activation of nuclear factor kappa B in the cochlea by sensory and inflammatory stress. Neuroscience 160(2):530–539

55. Vabulas RM, Ahmad-Nejad P, da Costa C, Miethke T, Kirschning CJ, Hacker H et al (2001) Endocytosed HSP60s use toll-like receptor 2 (TLR2) and TLR4 to activate the toll/interleu-kin-1 receptor signaling pathway in innate immune cells. J Biol Chem 276(33):31332–31339

56. Termeer C, Benedix F, Sleeman J, Fieber C, Voith U, Ahrens T et al (2002) Oligosaccharides of Hyaluronan activate dendritic cells via toll-like receptor 4. J Exp Med 195(1):99–111

57. Ohashi K, Burkart V, Flohe S, Kolb H (2000) Cutting edge: heat shock protein 60 is a putative endogenous ligand of the toll-like receptor-4 complex. J Immunol 164(2):558–561

58. Miyake K (2007) Innate immune sensing of pathogens and danger signals by cell surface Toll-like receptors. Semin Immunol 19(1):3–10

59. Zhang G, Ghosh S (2001) Toll-like receptor-mediated NF-kappaB activation: a phylogeneti-cally conserved paradigm in innate immunity. J Clin Invest 107(1):13–19

60. Kawai T, Akira S (2007) Signaling to NF-kappaB by Toll-like receptors. Trends Mol Med 13(11):460–469

61. Martin L, Pingle SC, Hallam DM, Rybak LP, Ramkumar V (2006) Activation of the adenos-ine A3 receptor in RAW 264.7 cells inhibits lipopolysaccharide-stimulated tumor necrosis

factor-alpha release by reducing calcium-dependent activation of nuclear factor-kappaB and extracellular signal-regulated kinase 1/2. J Pharmacol Exp Therap 316(1):71–78

62. Vethanayagam RR, Yang W, Dong Y, Hu BH (2016) Toll-like receptor 4 modulates the cochlear immune response to acoustic injury. Cell Death Dis 7(6):e2245

63. Oh GS, Kim HJ, Choi JH, Shen A, Kim CH, Kim SJ et al (2011) Activation of lipopolysaccharide-TLR4 signaling accelerates the ototoxic potential of cisplatin in mice. J Immunol 186(2):1140–1150

64. Xu Y, Chen S, Cao Y, Zhou P, Chen Z, Cheng K (2018) Discovery of novel small molecule TLR4 inhibitors as potent anti-inflammatory agents. Eur J Med Chem 154:253–266

65. Keithley EM, Wang X, Barkdull GC (2008) Tumor necrosis factor alpha can induce recruitment of inflammatory cells to the cochlea. Otol Neurotol 29(6):854–859

66. Wang X, Truong T, Billings PB, Harris JP, Keithley EM (2003) Blockage of immune-mediated inner ear damage by etanercept. Otol Neurotol 24(1):52–57

67. Nakamoto T, Mikuriya T, Sugahara K, Hirose Y, Hashimoto T, Shimogori H et al (2012) Geranylgeranylacetone suppresses noise-induced expression of proinflammatory cytokines in the cochlea. Auris Nasus Larynx 39(3):270–274

68. Lu B, Rutledge BJ, Gu L, Fiorillo J, Lukacs NW, Kunkel SL et al (1998) Abnormalities in monocyte recruitment and cytokine expression in monocyte chemoattractant protein 1-deficient mice. J Exp Med 187(4):601–608

69. Frye MD, Zhang C, Hu BH (2018) Lower level noise exposure that produces only TTS modulates the immune homeostasis of cochlear macrophages. J Neuroimmunol 323:152–166

70. Fredelius L, Rask-Andersen H, Johansson B, Urquiza R, Bagger-Sjoback D, Wersall J (1988) Time sequence of degeneration pattern of the organ of Corti after acoustic overstimulation. A light microscopical and electrophysiological investigation in the guinea pig. Acta Otolaryngol 106(1–2):81–93

71. Geissmann F, Jung S, Littman DR (2003) Blood monocytes consist of two principal subsets with distinct migratory properties. Immunity 19(1):71–82

72. Ziegler-Heitbrock L, Ancuta P, Crowe S, Dalod M, Grau V, Hart DN et al (2010) Nomenclature of monocytes and dendritic cells in blood. Blood 116(16):e74–e80

73. Auffray C, Fogg D, Garfa M, Elain G, Join-Lambert O, Kayal S et al (2007) Monitoring of blood vessels and tissues by a population of monocytes with patrolling behavior. Science 317(5838):666–670

74. Carlin LM, Stamatiades EG, Auffray C, Hanna RN, Glover L, Vizcay-Barrena G et al (2013) Nr4a1-dependent Ly6Clow monocytes monitor endothelial cells and orchestrate their disposal. Cell 153(2):362–375

75. Gilroy DW, Colville-Nash PR, McMaster S, Sawatzky DA, Willoughby DA, Lawrence T (2003) Inducible cyclooxygenase-derived 15-deoxy(Delta)12-14PGJ2 brings about acute inflammatory resolution in rat pleurisy by inducing neutrophil and macrophage apoptosis. FASEB J 17(15):2269–2271

76. Janssen WJ, Barthel L, Muldrow A, Oberley-Deegan RE, Kearns MT, Jakubzick C et al (2011) Fas determines differential fates of resident and recruited macrophages during resolution of acute lung injury. Am J Respir Crit Care Med 184(5):547–560

77. Hughes J, Johnson RJ, Mooney A, Hugo C, Gordon K, Savill J (1997) Neutrophil fate in experimental glomerular capillary injury in the rat. Emigration exceeds in situ clearance by apoptosis. Am J Pathol 150(1):223–234

78. Bellingan GJ, Caldwell H, Howie S, Dransfield I, Haslett C (1996) In vivo fate of the inflammatory macrophage during the resolution of inflammation: inflammatory macrophages do not die locally, but emigrate to the draining lymph nodes. J Immunol 157(6):2577–2585

79. Yimtae K, Song H, Billings P, Harris JP, Keithley EM (2001) Connection between the inner ear and the lymphatic system. Laryngoscope 111(9):1631–1635

80. Louveau A, Smirnov I, Keyes TJ, Eccles JD, Rouhani SJ, Peske JD et al (2015) Structural and functional features of central nervous system lymphatic vessels. Nature 523(7560):337

81. McWhorter FY, Wang T, Nguyen P, Chung T, Liu WF (2013) Modulation of macrophage phenotype by cell shape. Proc Natl Acad Sci U S A 110(43):17253–17258
82. Davis GS, Brody AR, Adler KB (1979) Functional and physiologic correlates of human alveolar macrophage cell shape and surface morphology. Chest 75(2 Suppl):280–282
83. Streit WJ, Graeber MB, Kreutzberg GW (1988) Functional plasticity of microglia: a review. Glia 1(5):301–307
84. Calton MA, Lee D, Sundaresan S, Mendus D, Leu R, Wangsawihardja F et al (2014) A lack of immune system genes causes loss in high frequency hearing but does not disrupt cochlear synapse maturation in mice. PLoS One 9(5):e94549
85. Fredelius L (1988) Time sequence of degeneration pattern of the organ of Corti after acoustic overstimulation. A transmission electron microscopy study. Acta Otolaryngol 106(5–6):373–385
86. Kaur T, Zamani D, Tong L, Rubel EW, Ohlemiller KK, Hirose K et al (2015) Fractalkine signaling regulates macrophage recruitment into the cochlea and promotes the survival of spiral ganglion neurons after selective hair cell lesion. J Neurosci. 35(45):15050–15061
87. Abrashkin KA, Izumikawa M, Miyazawa T, Wang CH, Crumling MA, Swiderski DL et al (2006) The fate of outer hair cells after acoustic or ototoxic insults. Hear Res 218(1–2):20–29
88. Sato E, Shick HE, Ransohoff RM, Hirose K (2008) Repopulation of cochlear macrophages in murine hematopoietic progenitor cell chimeras: the role of CX3CR1. J Comp Neurol 506(6):930–942
89. Lang H, Ebihara Y, Schmiedt RA, Minamiguchi H, Zhou D, Smythe N et al (2006) Contribution of bone marrow hematopoietic stem cells to adult mouse inner ear: mesenchymal cells and fibrocytes. J Comp Neurol 496(2):187–201
90. Bas E, Goncalves S, Adams M, Dinh CT, Bas JM, Van De Water TR et al (2015) Spiral ganglion cells and macrophages initiate neuro-inflammation and scarring following cochlear implantation. Front Cell Neurosci 9:303
91. Ruitenberg MJ, Vukovic J, Blomster L, Hall JM, Jung S, Filgueira L et al (2008) CX3CL1/fractalkine regulates branching and migration of monocyte-derived cells in the mouse olfactory epithelium. J Neuroimmunol 205(1–2):80–85
92. Jacquelin S, Licata F, Dorgham K, Hermand P, Poupel L, Guyon E et al (2013) CX3CR1 reduces Ly6Chigh-monocyte motility within and release from the bone marrow after chemotherapy in mice. Blood 122(5):674–683
93. Zhang J-M, An J (2007) Cytokines, inflammation and pain. Int Anesthesiol Clin 45(2):27
94. Zhang J-M, Li H, Liu B, Brull SJ (2002) Acute topical application of tumor necrosis factor α evokes protein kinase A-dependent responses in rat sensory neurons. J Neurophysiol 88(3):1387–1392
95. Özaktay AC, Kallakuri S, Takebayashi T, Cavanaugh JM, Asik I, DeLeo JA et al (2006) Effects of interleukin-1 beta, interleukin-6, and tumor necrosis factor on sensitivity of dorsal root ganglion and peripheral receptive fields in rats. Eur Spine J 15(10):1529–1537
96. Liu C, Glowatzki E, Fuchs PA (2015) Unmyelinated type II afferent neurons report cochlear damage. Proc Natl Acad Sci 112(47):14723–14727
97. Perry VH, Holmes C (2014) Microglial priming in neurodegenerative disease. Nat Rev Neurol 10(4):217–224
98. Hashimoto S, Billings P, Harris JP, Firestein GS, Keithley EM (2005) Innate immunity contributes to cochlear adaptive immune responses. Audiol Neurootol 10(1):35–43
99. Harris KC, Bielefeld E, Hu BH, Henderson D (2006) Increased resistance to free radical damage induced by low-level sound conditioning. Hear Res 213(1–2):118–129
100. Campo P, Subramaniam M, Henderson D (1991) The effect of 'conditioning' exposures on hearing loss from traumatic exposure. Hear Res 55(2):195–200
101. Subramaniam M, Henderson D, Campo P, Spongr V (1992) The effect of 'conditioning' on hearing loss from a high frequency traumatic exposure. Hear Res 58(1):57–62
102. Subramaniam M, Henderson D, Spongr V (1993) Effect of low-frequency "conditioning" on hearing loss from high-frequency exposure. J Acoust Soc Am 93(2):952–956

103. Subramaniam M, Henderson D, Spongr VP (1993) Protection from noise induced hearing loss: is prolonged 'conditioning' necessary? Hear Res 65(1–2):234–239
104. Henselman LW, Henderson D, Subramaniam M, Sallustio V (1994) The effect of 'conditioning' exposures on hearing loss from impulse noise. Hear Res 78(1):1–10
105. Hu BH, Henderson D (1997) Changes in F-actin labeling in the outer hair cell and the Deiters cell in the chinchilla cochlea following noise exposure. Hear Res 110(1–2):209–218
106. Roy S, Ryals MM, Van den Bruele AB, Fitzgerald TS, Cunningham LL (2013) Sound preconditioning therapy inhibits ototoxic hearing loss in mice. J Clin Invest 123(11):4945–4949
107. Perez R, Freeman S, Sohmer H (2004) Effect of an initial noise induced hearing loss on subsequent noise induced hearing loss. Hear Res 192(1–2):101–106
108. Zhang C, Frye MD, Sun W, Hu BH (2018) Preconditioning noise alters immune reaction to subsequent acoustic overstimulation in the cochlea. In: 41st Annual midwinter meeting, San Diego, CA
109. Rarey KE, Curtis LM, Wouter J-F (1993) Tissue specific levels of glucocorticoid receptor within the rat inner ear. Hear Res 64(2):205–210
110. ten Cate WJ, Curtis LM, Rarey KE (1992) Immunochemical detection of glucocorticoid receptors within rat cochlear and vestibular tissues. Hear Res 60(2):199–204
111. Takumi Y, Nishio S-Y, Mugridge K, Oguchi T, Hashimoto S, Suzuki N et al (2014) Gene expression pattern after insertion of dexamethasone-eluting electrode into the guinea pig cochlea. PLoS One 9(10):e110238
112. Lyu AR, Kim DH, Lee SH, Shin DS, Shin SA, Park YH (2018) Effects of dexamethasone on intracochlear inflammation and residual hearing after cochleostomy: a comparison of administration routes. PLoS One 13(3):e0195230
113. Chandrasekhar SS, Rubinstein RY, Kwartler JA, Gatz M, Connelly PE, Huang E et al (2000) Dexamethasone pharmacokinetics in the inner ear: comparison of route of administration and use of facilitating agents. Otolaryngol Head Neck Surg 122(4):521–528
114. Nanda S, Bathon JM (2004) Etanercept: a clinical review of current and emerging indications. Expert Opin Pharmacother 5(5):1175–1186
115. Rahman MU, Poe DS, Choi HK (2001) Etanercept therapy for immune-mediated cochleovestibular disorders: preliminary results in a pilot study. Otol Neurotol 22(5):619–624
116. Street I, Jobanputra P, Proops D (2006) Etanercept, a tumour necrosis factor α receptor antagonist, and methotrexate in acute sensorineural hearing loss. J Laryngol Otol 120(12):1064–1066
117. Matteson EL, Choi HK, Poe DS, Wise C, Lowe VJ, McDonald TJ et al (2005) Etanercept therapy for immune-mediated cochleovestibular disorders: a multi-center, open-label, pilot study. Arthritis Care Res 53(3):337–342

Preclinical Animal Behavioral Models of Hyperacusis and Loudness Recruitment

Kelly E. Radziwon, Senthilvelan Manohar, Benjamin Auerbach,
Xiaopeng Liu, Guang-Di Chen, and Richard Salvi

1 Introduction

1.1 Loudness

Acoustic stimuli give rise to subjective sensations, which the listener uses to interpret events occurring in the external environment, information crucial in making behaviorally relevant responses (e.g., reaction to a fire alarm). The subjective perception of frequency, measured in Hz (cycles per second), is referred to as pitch. Pitch increases with frequency within a biologically defined range. For humans, the frequency range of hearing extends from roughly 20 to 20,000 Hz, whereas for rats and mice typically used in research the frequency range is shifted upwards by several octaves. Loudness and abnormal loudness perception, the main topic of this chapter, is the subjective dimension of sound most closely related to sound pressure level or intensity [1]. As sound pressure level rises above the threshold of hearing (approximately 0 dB sound pressure level (SPL) at 4 kHz), loudness increases until the stimulus becomes intolerably loud or painful at intensities around 130–140 dB SPL [2–4]. The perceived loudness of a sound varies with not only sound pressure level but also a number of other acoustic variables.

For example, the loudness of a sound increases with increasing stimulus duration from approximately 1 ms up to 200–300 ms, a phenomenon referred to as temporal integration of loudness [5]. In addition, the perceived loudness of a sound composed of many frequencies remains constant when the spectral components of the sound are located near one another within the subject's critical bandwidth (i.e., spectral integration). However, sounds of equal overall intensity are perceived as louder as

K. E. Radziwon (✉) · S. Manohar · B. Auerbach · X. Liu · G.-D. Chen · R. Salvi
Center for Hearing and Deafness, University at Buffalo, Buffalo, NY, USA
e-mail: radziwon@buffalo.edu

© Springer Nature Switzerland AG 2020

S. Pucheu et al. (eds.), *New Therapies to Prevent or Cure Auditory Disorders*,
https://doi.org/10.1007/978-3-030-40413-0_6

the spectral components of the signal spread out and fall outside the critical band [6]. Furthermore, sounds presented to both ears are perceived as roughly twice as loud as a sound presented to one ear, a phenomenon referred to as binaural loudness summation [7].

Because the sensitivity of the ear varies with frequency, the loudness of a sound varies as a function of both intensity and frequency—reflected perceptually in measurements known as equal loudness contours. To generate equal-loudness contour plots, listeners are presented with sounds of different frequencies and sound pressure levels and asked to rate whether the sounds were equally loud. At low intensities, equal loudness contours have a dip near 3–4 kHz in humans, but the intensity rises at lower and higher frequencies in order to maintain equal loudness. However, equal-loudness contours flatten out at high intensities because loudness grows more rapidly with sound intensity at low and high frequencies [8]. In addition to frequency, loudness growth is also affected by background noise. Detection thresholds for tones increase with the addition of a masker or background noise. For intensities just above threshold, the perceived loudness of the tone in the presence of a background noise is reduced compared to the tone in quiet. However, as the intensity of the tone increases, loudness initially grows at a faster than normal rate until the loudness growth function catches up to the unmasked loudness growth function. Once this occurs, loudness growth in the masked ear slows and follows the normal growth of loudness [9, 10]. Loudness growth functions with a steeper than normal slope at low and moderate intensities, followed by normal growth and normal loudness at high intensities are referred to as recruitment-like.

1.2   Hearing Loss and Loudness Recruitment

Cochlear hearing loss due to aging, noise exposure, and ototoxic drugs not only raises the threshold for hearing but also alters loudness growth. In most cases of cochlear hearing loss, the loudness of a sound presented just above the threshold for hearing is perceived as softer than normal (less loud), but as the intensity of the stimulus is increased, loudness grows at a faster than normal rate so that the perceived loudness of a moderately intense sound eventually catches up and matches that of a normal listener. At higher intensities, loudness growth and loudness perception are the same as in a normal-hearing listener [11]. This two-stage loudness growth function in which loudness reaches normal levels at high intensities was described by Fowler nearly a century ago, who later coined the phrase loudness recruitment [12–14]. In some cases, softness imperception may also occur whereby the softest sound audible to listeners with sensorineural hearing loss is more intense than the softest sound audible to normal-hearing listeners [11].

1.3   Hyperacusis

Hyperacusis is an auditory hypersensitivity disorder in which every day sounds of moderate intensity are perceived as intolerably loud or even painful [15–21]. Loudness growth functions in hyperacusis patients are not only steeper than normal, but sounds presented at moderate to high intensities are perceived as much louder than normal. The mean loudness discomfort level (LDL) for normal-hearing subjects ranges from approximately 101 to 104 dB (SD: 11–14 dB) for frequencies from 0.5 to 4 kHz [22] whereas the mean LDL for hyperacusis patients is significantly lower with values ranging from 76 dB HL [23] to 85 dB HL [24].

Hyperacusis has a point prevalence of 9% [21, 25]. Among musicians, hyperacusis prevalence rates are as high as 60–70% [26, 27]. Hyperacusis seldom occurs in isolation; more often than not it is accompanied by hearing loss and tinnitus [18, 26, 28], disorders linked to cochlear damage induced by ototoxic drugs, noise exposure or aging [23, 29] as well as vestibular dysfunction [30]. Hyperacusis can occur in both ears or just one ear [31, 32]. Among patients with the primary complaint of tinnitus, approximately 40% also report hyperacusis [16, 23]. However, among those with a primary complaint of hyperacusis, up to 86% report that they also have tinnitus. While some individuals with hyperacusis appear to have clinically normal hearing, more detailed assessments reveal hearing loss at the extended high frequencies. In other cases, hearing thresholds can be normal despite extensive damage to the synapse between the inner hair cell and afferent auditory nerve fiber [21, 33–36], a condition referred to as hidden hearing loss because it is not detected by the standard pure tone audiogram.

1.4   Hyperacusis Phenotypes

A recent comprehensive review suggested there are four distinct types of hyperacusis based on a subject's loudness perception and reaction to sound, namely: (1) loudness hyperacusis, (2) annoyance hyperacusis, (3) fear hyperacusis, and (4) pain hyperacusis [37]. For a hearing-impaired subject with loudness hyperacusis, a low-intensity sound would be perceived as less loud than normal whereas a high-intensity sound would be perceived as louder than normal. Thus, individuals with loudness hyperacusis have lower than normal LDL values [24]. Some studies based on LDL suggest that loudness hyperacusis occurs at all frequencies, although these conclusions are based on limited data. In addition, while some studies suggest that loudness hyperacusis can occur unilaterally, others find that it occurs in both ears [15, 32, 38, 39]. Bilateral loudness hyperacusis, particularly with little or no cochlear pathophysiology, would be suggestive of a central dysfunction [29].

Annoyance hyperacusis refers to a condition in which a subject has negative emotional reactions to certain types of sound (e.g., chalk on a blackboard), leading to an increase in anxiety, annoyance, or tension [40–42]. Annoyance hyperacusis may arise from the unique acoustic features of a given sound that may activate specific regions of the brain, such as the amygdala, that assign emotional valence to sensory stimuli [43, 44]. Separate from annoyance, fear/avoidance hyperacusis reflects an anticipatory response aimed at escaping from a sound or class of acoustic stimuli that the subject perceives as aversive. In such cases, the subject engages in behaviors aimed at avoiding exposure to these sounds such as wearing ear plugs or staying away from situations where loud or aversive sounds might occur such as a concert or noisy restaurant.

Pain hyperacusis is a condition in which sounds are not only perceived as loud and aversive but also painful. For normal-hearing subjects, the threshold for evoking auditory pain is around 125–135 dB SPL [3, 4]. Subjects with pain hyperacusis experience painful or burning sensations within and around their ears at much lower sound intensities [45, 46]. Some speculate that pain hyperacusis could result from tissue damage to the cochlea. In turn, this information is then transmitted by the nonmyelinated type II auditory nerve fibers to the central nervous system [47, 48]. Others hypothesize that loud sounds activate pain receptors on the tympanic membrane or middle ear [3, 46]. Alternatively, loud sounds could activate neural networks involved in central pain [49–51].

1.5   Comorbidities

Hyperacusis is not just a hearing problem. Hyperacusis is linked to myriad genetic, neurological, and medical conditions such as autism, mental distress, vertigo, multiple sclerosis, Williams syndrome, migraine, cutaneous allodynia, superior canal dehiscence, multiple sclerosis, and fibromyalgia [17, 29, 30, 38, 52–57]. Hyperacusis is also associated with concentration difficulties, tension, sleep disturbances, anxiety, depression, cognitive impairments, and sensitivity to light/color [18, 58–64]. Taken together, these results suggest that hyperacusis is a more widespread and disruptive condition than currently recognized, one likely to involve not only the cochlea and other peripheral auditory structures, but other regions in the central nervous system as well.

1.6   Spectral Profile of Loudness Recruitment and Hyperacusis

There is an extensive literature on loudness recruitment in patients with cochlear hearing loss. Loudness recruitment consistently occurs in the region of hearing loss, but not in regions where hearing is normal [14, 65]. Thus, loudness recruitment has a spectral profile that is correlated with regions of hearing loss. Although hyperacusis is a well-recognized clinical problem, there is a dearth of information regarding the frequencies at which hyperacusis occurs, i.e., its spectral profile. One laboratory study from the 1950s found evidence of noise-induced hyperacusis at some frequencies, but loudness recruitment at other frequencies [66], raising the possibility of a spectral profile for hyperacusis. However, the results from a large retrospective study of patients suggest that LDLs are decreased across the full range of audiometric frequencies suggesting that hyperacusis occurs over a broad range of frequencies consistent with the notion of a generalized increased central gain or hyperexcitability [24].

2 Assessing Loudness Recruitment and Hyperacusis in Animal Models

The importance of animal behavioral models in hyperacusis research cannot be overstated [67, 68]. For one, a number of crucial electrophysiological, anatomical, and immunohistochemical assays for examining the underlying biological and physiological mechanisms of hyperacusis cannot be performed ethically in humans. Furthermore, hyperacusis is a complex disorder that cannot be studied in cell culture experiments alone. Hyperacusis involves altered auditory perception, disrupted central and peripheral pain mechanisms, and has emotional and memory components necessitating extensive behavioral and physiological examination. In addition, animal models are essential in the development of novel therapeutics to treat the disorder and rigorous testing of current hypotheses that cannot be performed directly in humans (e.g., type II neuronal function, enhanced central gain).

However, loudness recruitment and hyperacusis are subjective measures that are difficult to measure directly in animals. To assess the subjective nature of these clinical disorders, human listeners are often asked to assign a number or provide a written or verbal description to indicate how loud, annoying, or disruptive a sound is perceived [69–71]. In routine clinical practice, patients are usually asked to indicate the sound intensity that they perceive as being uncomfortably loud (uncomfortable loudness level (UCL) or loudness discomfort level (LDL)) [22, 24]. In other cases, subjects are asked to match the loudness of a sound presented to their normal ear to the loudness of the same or different sound presented to the contralateral hearing-impaired ear, a technique known as alternate binaural loudness balance test [72]. Loudness matching of stimuli can also be done in the same ear, in which case it is referred to as the monaural loudness balance test [73]. Hyperacusis questionnaires have also been developed to gather more in-depth information about the emotional, cognitive, and disruptive nature of acoustic stimuli [74–77]. While surveys, questionnaires, and loudness matching techniques can be readily employed with human listeners, these measurements cannot be obtained from animals in preclinical studies.

To overcome these limitations, researchers have developed animal behavioral models aimed at indirectly estimating how loud, annoying, or aversive a sound is perceived. These behavioral tests designed to assess loudness intolerance, sound aversion, annoyance, fear, and auditory pain provide a platform for testing drugs and other therapeutic techniques used to ameliorate or suppress the debilitating aspects of hyperacusis. In addition, these behavioral models can be used to identify drugs, genetic mutations, acoustic conditions, or neurological disorders that give rise to various aspects of hyperacusis [78–80]. Below, we describe several behavioral techniques that have been developed to assess loudness recruitment, loudness hyperacusis, sound aversion, and auditory pain in animal models.

2.1   Acoustic Startle Reflex

The acoustic startle reflex (ASR) is an abrupt, short-latency motoric response evoked by moderate- to high-intensity sounds. The ASR is primarily mediated by a brainstem circuit that includes the auditory nerve, posteroventral cochlea nucleus, ventrolateral lemniscus, caudal pontine reticular nucleus, and spinal motor neurons; however, the response can be modulated by higher level brain regions such as the inferior colliculus, amygdala, cerebellum, substantia nigra, and periaqueductal gray [81–88]. The amplitude of the ASR increases with sound intensity from approximately 65 to 115 dB SPL, but the response saturates at higher intensities (~50 dB dynamic range) [89]. Because the amplitude of the ASR increases with sound intensity, its amplitude is assumed to reflect, at least indirectly, the loudness of a sound. Hyperacusis is assumed to be present if ASR amplitudes are significantly larger than those measured during baseline testing or larger than those measured in a control group. The ASR has become a popular tool for making inferences about the presence of hyperacusis because it does not require any behavioral training and is relatively easy to measure.

High-dose sodium salicylate, the active ingredient in aspirin, induces temporary hearing loss and tinnitus in both humans and animals [90, 91]. Because tinnitus is often accompanied by hyperacusis, high-dose salicylate could conceivably alter the ASR. To test this hypothesis, ASR input/output functions were measured before and after high-dose salicylate treatment. High-dose salicylate, but not saline, significantly increased ASR amplitudes [92, 93], results consistent with an early clinical report indicating that salicylate induces hyperacusis [94]. The increase in ASR amplitude was associated with sound-evoked hyperactivity in the auditory cortex, medial geniculate body [92] and importantly, the caudal pontine reticular nucleus, a critical part of the ASR circuit [95].

Others have reported increases in ASR amplitude in animals with age-related high-frequency hearing loss [96, 97]; these results were also associated with sound-evoked hyperexcitability in the inferior colliculus. The effects of noise exposure on ASR amplitude have varied. In some cases, noise exposure caused transient or chronic increases in startle amplitude [98–100]. Changes in ASR amplitude following intense, short-duration noise exposures depend on many factors [101, 102]. ASR amplitudes were typically enhanced with mild hearing loss, depressed with large hearing loss, and unaffected by little or no hearing loss. Furthermore, the time course of ASR amplitude changes tended to be depressed the first week after the exposure, but became enhanced several weeks later [101]. Interestingly, ASR amplitudes in response to the startle stimulus were not enhanced in noise-exposed animals when responses were measured in a gap detection pre-pulse paradigm, i.e., the acoustic context affected ASR amplitude enhancement.

The extent to which ASR amplitude can be used to infer hyperacusis is currently unclear. In a study involving humans with nearly normal thresholds, eyeblink ASR amplitudes were larger in those with lower LDL values (i.e., ASR negatively correlated LDL); however, LDL values were not correlated with hyperacusis test scores [103] as previously noted [23, 104]. These results suggested that clinical LDL measures do not accurately reflect real-life loudness judgments. Because the ASR is primarily driven by neurons in the lower brainstem, the reflex response may be more indicative of pre-attentive neural activity than higher level perceptual or cognitive processes. Taken together, these results suggest that the ASR should be used with caution when making inferences about hyperacusis in animal models.

2.2   Reaction Time-Intensity Functions and Loudness Growth

A well-established metric for assessing loudness in humans and animals is auditory reaction time (RT), i.e., the time between the onset of a sound and the time it takes for the listener to respond. RT is closely related to loudness perception across a wide range of stimulus levels and frequencies [105, 106] for both normal-hearing [107, 108] and hearing-impaired listeners [109]. In both human and animal studies, there is an inverse relationship between RT and stimulus intensity, i.e., *RT decreases with increasing intensity* [107, 108, 110–114]. Since RT is a well-established correlate of loudness, RT-intensity (RT-I) functions can accurately measure loudness growth in both humans and animals. Importantly, detailed RT-I functions can separate listeners with normal loudness growth from those experiencing loudness recruitment or hyperacusis (Fig. 1a). Loudness recruitment, often associated with cochlear hearing loss, is defined as a steeper than normal growth of loudness at intensities just above threshold, with normal loudness growth and normal loudness at high sound intensities. In contrast, hyperacusis is characterized by a steeper growth of loudness at both moderate and high sound intensities, resulting in moderate- to high-intensity sounds being perceived as louder than normal. Thus, individuals with hyperacusis typically have faster-than-normal RTs at moderate to high intensities (Fig. 1a, red line) whereas those with loudness recruitment have RTs that are normal at moderate and high intensities but slower at low levels (Fig. 1a, green line).

To obtain RT-I functions, animals are trained using a Go/No-go operant conditioning paradigm. Generally, animals are trained to make a response (e.g., press a lever, or poke their noses into a hole) when they detect a sound (*Go*) while withholding their responses when they do not hear a sound (*No-go*). Using this simple behavioral response, researchers can simultaneously determine the animal's threshold of hearing to obtain an audiogram while collecting RTs to suprathreshold sounds to obtain detailed RT-I loudness growth functions.

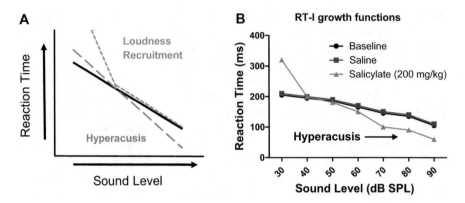

Fig. 1 (**a**) Schematic represenation of reaction time–intensity (RT-I) functions for listeners with normal loudess growth (black line), loudness recruitment (green small-dashed line), and hyperacusis (red large-dashed line). (**b**) RT-I functions for rats before salicylate administration (baseline; black circles), after a control saline injection (Saline; blue squares), and 2 h post-salicylate administration (salicylate; red triangles). High-dose salicyate induced slower than normal RTs for 30 dB SPL noisebursts, indicative of temporary hearing loss, and faster than normal RTs for moderate- to high-intensity sounds, behavior indicative of loudness hyperacusis. (Schematic of data from Chen et al. 2014)

RT-I functions of loudness growth follow the psychophysical rules of loudness perception. RT is faster for broadband noise stimuli than tone bursts, consistent with the spectral integration of loudness seen in human psychophysical studies [6, 115]. In addition, RT decreases with stimulus duration consistent with temporal integration of loudness seen in psychophysical studies [7, 116]. In the presence of background noise, RT-I functions show evidence of loudness recruitment [117] similar to that observed in human psychophysical studies of loudness growth [10, 118]. While the RT-I operant conditioning method produces a reliable and extremely robust measure of loudness growth across a wide dynamic range in both human and animal listeners, this method is labor-intensive and time-consuming.

Despite its drawbacks, many investigators have found evidence of loudness recruitment in various animal models of noise-induced and drug-induced hearing loss [110, 111, 119, 120], making the RT-I conditioning method an extremely useful technique for assessing normal and abnormal loudness growth. However, few studies have seen faster than normal RTs indicative of hyperacusis. In one of the earliest reports, Lauer and Dooling found faster than normal RTs at moderate and high intensities in genetically hearing-impaired canaries. Their findings were complemented by data from a human subject with autism spectrum disorder who had faster than normal RTs [112], behavioral RT data consistent with prior observations of hyperacusis, and sensory hypersensitivity disorders in autism and autism spectrum disorder [52, 121].

As previously discussed, high doses of sodium salicylate reliably induce tinnitus and temporary hearing loss of approximately 20–25 dB across all frequencies in both humans and animals [90, 91, 122, 123]. One very early study mentioned that high-dose salicylate also induces hyperacusis in humans [94], but we are unaware of any other studies confirming this. To determine if salicylate could induce hyperacusis, Radziwon and colleagues measured RT-I functions in rats before and after treating them with a high dose of sodium salicylate (200 mg/kg). Two hours following salicylate treatment, there was a striking reduction of RT to noise bursts presented at moderate to high intensities (Fig. 1b, red line). These behavioral results indicate that the rats were experiencing sounds above 50 dB SPL as louder than normal [93]. In contrast, RTs were slower than normal at 30 dB SPL due to the temporary hearing loss; therefore, at low intensities, the sounds were less loud than normal. Importantly, RT-I functions returned to normal when SS treatment was discontinued. Control injections with saline had no effect on RT-I functions (Fig. 1b, blue line). Since these RT measures were obtained from the same animals before, during, and after SS treatment, the results provide compelling evidence that salicylate-treated rats developed temporary hyperacusis [93, 115].

In a subsequent dose–response study, salicylate-induced hyperacusis occurred with doses of 150 mg/kg or higher, whereas RT-I was largely unaffected by doses equal to or less than 100 mg/kg. The doses that reliably induced hyperacusis were identical to those that reliably induced tinnitus reinforcing the notion that hyperacusis is often comorbid with tinnitus [91]. Interestingly, the doses of salicylate that induced hyperacusis and tinnitus also caused a large increase in corticosterone, a hormone linked to stress and anxiety [95], a finding with mechanistic implications. When RT-I functions were measured with tone bursts from 4 to 20 kHz, salicylate-induced hyperacusis was observed across all stimulus frequencies consistent with studies showing that salicylate induces a relatively flat hearing loss of 20–25 dB.

We recently measured RT-I functions before, during, and several months after exposing rats to an intense high-frequency noise (16–20 kHz) that induced approximately 60 dB of permanent high-frequency (>12 kHz) hearing loss while retaining normal low-frequency hearing. High-frequency RT-I functions showed evidence of loudness recruitment both during and after the noise exposure. In contrast, low-frequency RT-I functions revealed robust evidence of hyperacusis (faster RTs) after the noise exposure ended. The hyperacusis-like behavior persisted with RTs steadily decreasing from 1 to 3 months post-exposure. Thus, hyperacusis gradually developed over a period of several months after inducing a high-frequency hearing loss. Hyperacusis was located below and near the edge of the high-frequency hearing loss whereas loudness recruitment was present in the region of severe hearing loss.

2.3   Two-Alternative Forced Choice Loudness Categorization Paradigm

Similar to the RT-I behavioral model mentioned above, some investigators have used a two-alternative forced choice (2AFC) operant behavioral model in which animals are asked to categorize the loudness of sounds presented at different intensities [124, 125]. In this paradigm, a food-restricted rat is trained to initiate a trial by poking its nose into a nose-poke hole. The animal is trained to remove its nose from nose-poke hole when a sound is presented and make a categorical decision as to whether the sound is perceived as "loud" or "quiet." During the training period, the animal is taught to identify a high-intensity sound (100 dB, perceived as very loud) versus a low-intensity sound (60 dB, perceived as soft). If a low-intensity sound is presented, the animal received a food reward if it responds to the food pellet dispenser on the left. If a high-intensity sound was presented, the animal receives a food reward when it responds to the right food pellet dispenser. Training continued until the rat correctly categorized the high-intensity and low-intensity sounds at greater than 90% correct. Once the rat correctly categorized the endpoint stimuli (60 versus 100 dB SPL), probe stimuli were presented at randomly selected intensities, between 40 and 110 dB SPL. On the probe trials, the rat has to respond to the left or right and in so doing categorized the sound as belonging to the "soft" or "loud" sound category but there is no "correct" side for probe stimuli.

Using this approach, the rats generate a sigmoidal function, a plot which shows the percentage of times a sound of a given intensity is judged to belong to the "loud" sound category. Tone bursts around 95 dB SPL are judged as belonging to the "loud" sound category nearly 100% of the time, but as the intensity decreases, the percentage declines towards 0% (i.e., "soft"). The "loudness" versus intensity function determines how the rats categorize sound intensities along the 40–110 dB SPL continuum. An experimental manipulation that shifts the "percent loud" versus intensity function towards the left (low-intensity) would indicate that intermediate intensities are now perceived as louder than before [124, 126] whereas a shift of the function to the right was interpreted as evidence of loudness recruitment. Because the percentages near the high-intensity endpoint are often all nearly 100%, the "percent loud" versus intensity function is not very sensitive (i.e., insensitive) at detecting changes in loudness at high intensities where hyperacusis is most likely to occur. Thus, this procedure is most suited at detecting changes in loudness at intermediate intensities between the endpoints. As with other operant behavioral paradigms, this 2AFC loudness categorization model requires extensive training to obtain optimum task performance [127].

Using this 2AFC procedure, Sun and colleagues found that high-dose salicylate produced evidence of hyperacusis in approximately 40% of rats, while another 40% developed evidence of loudness recruitment [124]. The same procedure was used to test for changes in loudness perception in adult rats that had recovered from a conductive hearing loss (tympanic membrane damage) that occurred when they were still young pups [125]. Rats with early conductive hearing loss typically categorized sounds as "louder" than their age-matched counterparts. Interestingly, the rats with early conductive hearing loss also developed audiogenic seizures. Treatment with vigabatrin, an anti-seizure drug that inhibits the catabolism of GABA, an inhibitory neurotransmitter, not only prevented audiogenic seizures, but also attenuated the exaggerated loudness perception. Mechanistically, the behavioral phenotypes were associated with reduced expression of GABA receptor delta and alpha 6 subunits in the inferior colliculus. One interpretation of these results is that early conductive hearing loss, often experienced by children with a history of chronic otitis media, may predispose individuals to hyperacusis when they become adults. In line with this animal study, normal-hearing adults who wear ear plugs for a week or two, experience roughly a 7 dB, temporary reduction in their LDL values after the ear plugs are removed [128]. Four days of unilateral auditory deprivation has also been associated with a temporary reduction in acoustic reflex thresholds in the sound-deprived ear suggestive of an increase in central auditory gain [129]. However, wave V amplitude of the auditory brainstem response was reduced in the sound-deprived ear, but enhanced in the control ear, changes opposite to those predicted by central gain models of hyperacusis.

2.4 Conditioned Avoidance Paradigm

Rüttiger and colleagues originally developed their water-reinforced conditioned avoidance paradigm to assess tinnitus in animals [130]. The animals are trained to shuttle between two water spouts during the presentation of a background noise to receive a sugar water reward. However, the animals learn to avoid the water spouts when the background noise is turned off (conditioned avoidance) because if they lick during the silent interval, they receive a mild foot shock. After training, the foot shock is turned off and sounds ranging from 20 to 60 dB, along with quiet intervals, are presented to the animal. Following noise trauma, animals with robust drinking during the quiet intervals are believed to have tinnitus while animals with elevated drinking behavior during low-level sound trials are believed to have hyperacusis [131]. Although this paradigm has the advantage of measuring both tinnitus and hyperacusis in the same animal, the conditioned avoidance model could be problematic for repeated testing of chronic hyperacusis or tinnitus. The reason for this is that following repeated testing, the animals eventually learn that the foot shock is turned off and the animals will no longer be under stimulus control because the previously trained reinforcement rules have been removed [127].

2.5 Active Sound Avoidance Paradigm

The preceding behavioral techniques have mainly focused on loudness hyperacusis and loudness recruitment. However, some human listeners react to intense sounds by engaging in active sound avoidance behaviors because the acoustic stimuli evoke fear, anxiety, or annoyance [37]. These reactions to sound often result in sound avoidance behaviors such as wearing earplugs or avoiding potentially noisy work or social environments [23, 56, 132]. Human sound avoidance behaviors are typically assessed by surveys or questionnaires aimed at gathering subjective insights into a person's reaction to sound. While these methods cannot be applied to animals, one study has described a method to assess sound avoidance behaviors in normal-hearing and hearing-impaired rats.

Using a modified place preference test apparatus, sound avoidance was assessed in rats with an active sound avoidance paradigm (ASAP) [133]. ASAP is based on a rat's innate aversion of bright, open spaces and preference for dark enclosures. By placing a loudspeaker inside a dark, sound-attenuating box, sound stimuli can be presented at different intensities and frequencies to identify the acoustic conditions that cause a rat to shift it preference from the dark enclosure with sound into the bright, open arena. Because of the sound-attenuating properties of the dark enclosure, sound intensity in the bright, open space is considerably less than inside the enclosure (Fig. 2a). The sounds that cause the rat to shift its preference from the dark box to the light box presumably evoke annoyance, fear, or anxiety. ASAP sound avoidance behavior is assessed by measuring the percentage of time a rat spends in the dark box under ambient acoustic conditions and again when sounds of various intensities and spectral content are presented. If the animal leaves the dark box when a given sound is present, the shift in preference from the dark box to the bright open space can be used to quantify sound avoidance. Unlike the operant conditioning paradigms described above, ASAP does not require extensive training. Results suggest that ASAP assesses the emotional valence of sound perception that cannot be captured in the RT-I, or other operant procedures, which mainly reflect loudness perception [133].

To determine if high-frequency hearing loss would alter sound avoidance behavior, rats were tested on ASAP before and 2 weeks after being exposed to an intense high-frequency noise (16–20 kHz, 102 dB SPL, 6-week exposure). Before the noise exposure, the rats spent roughly 95% of the time in the dark box under ambient background noise (Quiet), but when a 2–8 kHz noise was introduced at 60 and

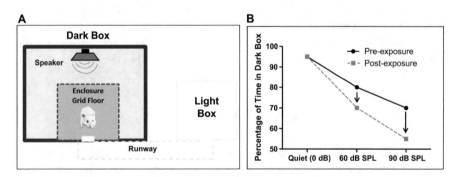

Fig. 2 (**a**) ASAP apparatus schematic. Sound-attenuating dark box contains a speaker and rat cage with a grid floor. The dark box is connected to an open, light box by a plexiglas runway. The rats can avoid sounds presented in the dark box by entering the light box. (**b**) Percent of time spent in the dark box in rats pre-noise exposure (black circles) and post-noise exposure (16–20 kHz, 102 dB SPL, 6 weeks; red squares) as a function of intensity. After inducing a permanent high-frequency hearing loss, the rats spent significantly less time in the dark box when both the 60 and 90 dB SPL sounds were present, indicative of avoidance hyperacusis. (Schematic of data from Manohar et al. 2017)

90 dB SPL, the rats spent progressively less time in the dark box, behavioral evidence of sound avoidance (Fig. 2b, black circles) [133]. After inducing a 40–50 dB high-frequency loss (>12 kHz), sound avoidance in the Quiet condition was unaffected; however, sound avoidance behavior significantly increased (reduced time in the dark box) when the 2–8 kHz noise, located in the region of normal hearing, was presented at either 60 or 90 dB SPL (Fig. 2b, red squares). Importantly, the noise-exposed rats showed no differences in their preference for the dark box in the Quiet condition. Thus, the high-frequency hearing loss did not cause an overall change in light/dark preference; rather, the noise exposure induced sound-evoked avoidance behavior [133]. While ASAP appears to be able to tap into the emotional aspects of sound perception, ASAP as currently configured is somewhat labor intensive and it is unclear if repeated testing causes the animals to habituate to the testing setup or stimulus conditions.

2.6 Multisensory Interaction Between Auditory and Pain Pathways

Some hyperacusis patients not only perceive high-intensity sounds as too loud or aversive but also experience painful sensations around the ear, face, and mouth [37, 134, 135]. Pain hyperacusis is a complex phenomenon with comorbid symptoms resembling complex regional pain syndrome [136], migraine [56], temporomandibular disorder [137], fibromyalgia [138], and other neurological syndromes thought to result from central pain sensitization. While animal models of pain typically utilize behavioral characterization of overall health and facial grimace markers [139, 140] as indices of pain, these metrics have been difficult to study in the context of sound-evoked pain and have not yet been investigated. However, Manohar and colleagues have investigated the effects of sound stimulation on thermal pain sensitivity in rats using a technique referred to as the auditory nociception test (ANT). Broadly speaking, ANT evaluates multisensory integration between the auditory and thermal pain pathways, an interaction that may occur in the central auditory pathway or elsewhere in the central nervous system.

ANT uses the time it takes for a rat to withdraw its tail (tail-flick latency) from warm water (52 °C) as a metric for assessing thermal pain sensitivity. It is a noninvasive and easy to repeat method for assessing multisensory integration between the auditory and thermal pain pathways, specifically the effect that sounds presented at different frequencies and intensities have on thermal pain sensitivity. A shorter withdrawal latency indicates greater pain sensitivity (i.e., hyperalgesia) whereas a longer latency reflects less thermal pain sensitivity (i.e., hypoalgesia). The tail-flick latency test is commonly used in nociceptive drug research [141, 142]. However, to apply the tail-flick latency test to studies involving multisensory auditory-pain integration, a loudspeaker is used to determine the effects of sound stimulation on thermal tail-flick latencies (Fig. 3a).

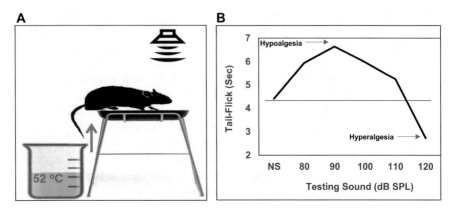

Fig. 3 (**a**) ANT schematic. Noise from overhead speaker is generated for 60 s. The rat's tail is dipped in the water after the sound is on for 45 s. Tail-flick latencies are measured during the last 15 s of sound presentation. (**b**) Tail-flick latencies for normal-hearing rats as a function of sound intensity. Tail-flick latency at ambient noise level ranges from 4 to 5 s (NS condition). When noise reaches 90 and 100 dB SPL, tail-flick latency increases to approximately 6–7 s indicative of reduced pain sensation (hypoalgesia), whereas tail-flick latency falls to 2–3 s at 120 dB SPL, indicative of enhanced pain sensation or hyperalgesia. (Schematic of unpublished data from S. Manohar)

Prior studies in humans and animals have shown that moderate-intensity sounds decrease pain (hypoalgesia) and therefore increase pain thresholds and tail withdrawal latencies, an effect known as the audio-analgesic effect [143–147]. Indeed, audio-analgesia, a non-pharmacologic approach, has been employed clinically for many years [148–154]. While there is an extensive literature related to audio-analgesia evoked by moderate-intensity sounds, little is known about the effects of intense sounds on pain sensitivity. Because intense sounds typically cause auditory pain [3], high-intensity sounds would presumably decrease tail-flick latencies due to the integration of auditory pain and thermal pain at locations in the central nervous system where these two sensory pathways converge [155–158]. To test this hypothesis, tail-flick latencies in normal-hearing rats were measured as the intensity of a broadband noise test sound was increased from ambient conditions to 120 dB SPL. Tail withdrawal latencies increased in the presence of moderately intense sounds (80–100 dB SPL) consistent with previous studies of audio-analgesia (Fig. 3b). However, at higher intensities near the auditory threshold of pain, tail-flick latencies decreased and were significantly below the latencies obtained during ambient sound conditions (Fig. 3b). These results indicate that high-intensity sounds, near the threshold of pain, exacerbate thermal pain (hyperalgesia, more painful), whereas moderately intense sounds reduce pain sensation. Taken together, these results provide compelling evidence for multisensory integration of sensory information from auditory and pain pathways. The implication of these behavioral findings for pain hyperacusis is unclear because thermal tail-pain nociception is transmitted by the caudal regions of the dorsal root ganglion, a region with unknown convergence with auditory centers. However, the convergence of pain information from the dorsal root ganglion and auditory regions seems plausible given that patients with whiplash and traumatic spinal cord injury sometimes develop tinnitus and sensory hypersensitivity [159–161].

3 Synopsis

Hyperacusis is an underdiagnosed sensory hypersensitivity disorder that often accompanies tinnitus [21]. Like many other sensory hypersensitivity disorders, hyperacusis is associated with a plethora of seemingly unrelated neurological conditions such as migraine, fibromyalgia, autism, and multiple sclerosis. However, it is often overlooked by patients and clinicians because patients are unaware of their condition due to its subjective qualities. Nevertheless, severe hyperacusis can be an extremely debilitating condition contributing to strong emotional reactions, dramatic lifestyle changes, and sound-evoked pain. The neural and biological mechanisms that give rise to hyperacusis are poorly understood. Nevertheless, there is a growing body of evidence suggesting that loudness intolerance disorders are associated with sound-evoked hyperactivity in both auditory and nonauditory portions of the central nervous system [162–165]. In order to develop a clearer understanding of the biological basis of hyperacusis, it is important to develop valid behavioral models that provide accurate readouts of hyperacusis associated with loudness, pain, and its emotional characteristics that mimics some of the phenotypes identified in humans. Among the various preclinical models, several behavioral paradigms could be used to investigate loudness hyperacusis. The ASR is one of the easiest and straightforward to implement. However, the ASR is only weakly correlated with LDL values and is uncorrelated with hyperacusis questionnaire scores. Moreover, ASR amplitudes are greatly diminished by hearing loss. The 2AFC loudness categorization method appears to be a useful method for identifying changes in loudness at intermediate intensities, but appears to be relatively insensitive to changes in loudness associated with very intense acoustic stimulation. While measurements of loudness growth obtained with RT-I functions take considerable time to collect, this measure obeys many of the rules of loudness growth seen in humans (e.g., temporal and spectral integration, masking). As such, RT-I functions appear to be one of the most valid measures of loudness growth, loudness recruitment, and hyperacusis that can also be implemented in an animal model. RT-I functions have proved useful in identifying hyperacusis in a canary model of a genetic cochlear hearing loss and in detecting hyperacusis in a human with autism spectrum disorder [112]. Moreover, RT-I functions have proved useful in identifying the spectral location of loudness recruitment versus hyperacusis in a model of noise-induced high-frequency hearing loss. Thus, the RT-I method could be used to identify hyperacusis in animal models with genetic disorders associated with sensory hypersensitivity such as Williams syndrome, early conductive hearing loss, or the types of noise exposures (blast wave, impulse noise, long-duration noise exposure) most likely to induce persistent hyperacusis.

Because hyperacusis is often associated with noise-induced hearing and genetic disorders such as autism, RT-I functions could also provide a valuable readout to screen for drugs to reverse or suppress various forms of hyperacusis. While progress has been made in developing the ANT behavioral model, which assesses the interaction between the auditory and pain pathways, it is unclear whether this model accurately reflects pain hyperacusis seen in humans who experience pain within and around the ear. Human pain hyperacusis is likely to involve pain relayed by portions of the trigeminal and facial nerve innervating the external, middle, and inner ear and areas surrounding the external ear and face [46].

Acknowledgments Supported in part by grant from NIH (R01DC014452, R01DC014693 and F32 DC015160), Hearing Health Foundation and Simmons Foundation.

References

1. Abur D, Lupiani AA, Hickox AE, Shinn-Cunningham BG, Stepp CE (2018) Loudness perception of pure tones in Parkinson's disease. J Speech Lang Hear Res 61(6):1487–1496
2. Franks JR, Stephenson MR, Merry CJ (1996) Preventing occupational hearing loss—a practical guide. In: Health NIfOSa (ed) DHHS (NIOSH)
3. Ades HW, Morrill SN, Graybiel A, Tolhurst GC (1959) Threshold of aural pain to high intensity sound. Aerosp Med 30:678–684
4. Gierke HEV, Davis H, Eldredge DH, Hardy JD (1953) Aural pain produced by sound. In: Benox report, an exploratory study of the biological effects of noise. Project NR 144079. Office of Naval Research, University of Chicago Press
5. Poulsen T (1981) Loudness of tone pulses in a free field. J Acoust Soc Am 69(6):1786–1790
6. Scharf B, Bands C (1970) In: Tobias J (ed) Foundations of modern auditory I. Academic Press, New York, pp 157–202
7. Algom D, Rubin A, Cohen-Raz L (1989) Binaural and temporal integration of the loudness of tones and noises. Percept Psychophys 46(2):155–166
8. Schneider BA, Bissett RJ (1987) Equal loudness contours derived from comparisons of sensory differences. Can J Psychol 41(4):429–441
9. Hellman RP (1994) Relation between the growth of loudness and high-frequency excitation. J Acoust Soc Am 96(5 Pt 1):2655–2663
10. Richards AM (1973) Loudness growth under masking: relation to true sensorineural impairment. J Speech Hear Res 16(4):597–607
11. Marozeau J, Florentine M (2007) Loudness growth in individual listeners with hearing losses: a review. J Acoust Soc Am 122(3):EL81
12. Fowler EP (1937) The diagnosis of diseases of the neural mechanism of hearing by the aid of sounds well above threshold. Laryngoscope (St Louis) 47:289–300
13. Fowler EP (1963) Loudness recruitment. Definition and clarification. Arch Otolaryngol 78:748–753
14. Dix MR, Hallpike CS, Hood JD (1948) Observations upon the loudness recruitment phenomenon, with especial reference to the differential diagnosis of disorders of the internal ear and 8th nerve. J Laryngol Otol 62(11):671–686
15. Valente M, Goebel J, Duddy D, Sinks B, Peterein J (2000) Evaluation and treatment of severe hyperacusis. J Am Acad Audiol 11(6):295–299
16. Baguley DM (2003) Hyperacusis. J R Soc Med 96(12):582–585

17. Schecklmann M, Landgrebe M, Langguth B, Group TRIDS (2014) Phenotypic characteristics of hyperacusis in tinnitus. PLoS One 9(1):e86944
18. Andersson G, Lindvall N, Hursti T, Carlbring P (2002) Hypersensitivity to sound (hyperacusis): a prevalence study conducted via the internet and post. Int J Audiol 41(8):545–554
19. Zeng FG (2013) An active loudness model suggesting tinnitus as increased central noise and hyperacusis as increased nonlinear gain. Hear Res 295:172–179
20. Westcott M (2006) Acoustic shock injury (ASI). Acta Otolaryngol Suppl 556:54–58
21. Gu JW, Halpin CF, Nam EC, Levine RA, Melcher JR (2010) Tinnitus, diminished sound-level tolerance, and elevated auditory activity in humans with clinically normal hearing sensitivity. J Neurophysiol 104(6):3361–3370
22. Sherlock LP, Formby C (2005) Estimates of loudness, loudness discomfort, and the auditory dynamic range: normative estimates, comparison of procedures, and test-retest reliability. J Am Acad Audiol 16(2):85–100
23. Anari M, Axelsson A, Eliasson A, Magnusson L (1999) Hypersensitivity to sound—questionnaire data, audiometry and classification. Scand Audiol 28(4):219–230
24. Sheldrake J, Diehl PU, Schaette R (2015) Audiometric characteristics of hyperacusis patients. Front Neurol 6:105
25. Goldstein B, Shulman A (1996) Tinnitus—hyperacusis and the loudness discomfort level test—a preliminary report. Int Tinnitus J 2:83–89
26. Schmuzigert N, Fostiropoulos K, Probst R (2006) Long-term assessment of auditory changes resulting from a single noise exposure associated with non-occupational activities. Int J Audiol 45(1):46–54
27. Kahari K, Zachau G, Eklof M, Sandsjo L, Moller C (2003) Assessment of hearing and hearing disorders in rock/jazz musicians. Int J Audiol 42(5):279–288
28. Laitinen H, Poulsen T (2008) Questionnaire investigation of musicians' use of hearing protectors, self reported hearing disorders, and their experience of their working environment. Int J Audiol 47(4):160–168
29. Weber H, Pfadenhauer K, Stohr M, Rosler A (2002) Central hyperacusis with phonophobia in multiple sclerosis. Mult Scler 8(6):505–509
30. Minor LB, Cremer PD, Carey JP, Della Santina CC, Streubel SO, Weg N (2001) Symptoms and signs in superior canal dehiscence syndrome. Ann N Y Acad Sci 942:259–273
31. Brookler KH (2009) Monaural diplacusis with tinnitus, aural fullness, hyperacusis, and sensorineural hearing loss. Ear Nose Throat J 88(2):772–774
32. Boucher O, Turgeon C, Champoux S, Menard L, Rouleau I, Lassonde M et al (2015) Hyperacusis following unilateral damage to the insular cortex: a three-case report. Brain Res 1606:102–112
33. Lin HW, Furman AC, Kujawa SG, Liberman MC (2011) Primary neural degeneration in the Guinea pig cochlea after reversible noise-induced threshold shift. J Assoc Res Otolaryngol 12(5):605–616
34. Sergeyenko Y, Lall K, Liberman MC, Kujawa SG (2013) Age-related cochlear synaptopathy: an early-onset contributor to auditory functional decline. J Neurosci 33(34):13686–13694
35. Weisz N, Hartmann T, Dohrmann K, Schlee W, Norena A (2006) High-frequency tinnitus without hearing loss does not mean absence of deafferentation. Hear Res 222(1–2):108–114
36. Kiani F, Yoganantha U, Tan CM, Meddis R, Schaette R (2013) Off-frequency listening in subjects with chronic tinnitus. Hear Res 306:1–10
37. Tyler RS, Pienkowski M, Roncancio ER, Jun HJ, Brozoski T, Dauman N et al (2014) A review of hyperacusis and future directions: part I. Definitions and manifestations. Am J Audiol 23(4):402–419
38. Dang PT, Kennedy TA, Gubbels SP (2014) Simultaneous, unilateral plugging of superior and posterior semicircular canal dehiscences to treat debilitating hyperacusis. J Laryngol Otol 128(2):174–178

39. Miani C, Passon P, Bracale AM, Barotti A, Panzolli N (2001) Treatment of hyperacusis in Williams syndrome with bilateral conductive hearing loss. Eur Arch Otorhinolaryngol 258(7):341–344
40. Gast PL, Baker CF (1989) The CCU patient: anxiety and annoyance to noise. Crit Care Nurs Q 12(3):39–54
41. Abel SM (1990) The extra-auditory effects of noise and annoyance: an overview of research. J Otolaryngol 19(Suppl 1):1–13
42. Urnau D, Tochetto TM (2011) Characteristics of the tinnitus and hyperacusis in normal hearing individuals. Int Arch Otorhinolaryngol 15:468–474
43. Kumar S, Tansley-Hancock O, Sedley W, Winston JS, Callaghan MF, Allen M et al (2017) The brain basis for misophonia. Curr Biol 27(4):527–533
44. Kumar S, von Kriegstein K, Friston K, Griffiths TD (2012) Features versus feelings: dissociable representations of the acoustic features and valence of aversive sounds. J Neurosci 32(41):14184–14192
45. Baguley DM, Hoare DJ (2018) Hyperacusis: major research questions. HNO 66(5):358–363
46. Norena AJ, Fournier P, Londero A, Ponsot D, Charpentier N (2018) An integrative model accounting for the symptom cluster triggered after an acoustic shock. Trends Hear 22:2331216518801725
47. Flores EN, Duggan A, Madathany T, Hogan AK, Marquez FG, Kumar G et al (2015) A non-canonical pathway from cochlea to brain signals tissue-damaging noise. Curr Biol 25(5):606–612
48. Liu C, Glowatzki E, Fuchs PA (2015) Unmyelinated type II afferent neurons report cochlear damage. Proc Natl Acad Sci U S A 112(47):14723–14727
49. Balaban CD (2011) Migraine, vertigo and migrainous vertigo: links between vestibular and pain mechanisms. J Vestib Res 21(6):315–321
50. Pfadenhauer K, Weber H, Rosler A, Stohr M (2001) Central hyperacusis with phonophobia in multiple sclerosis. Nervenarzt 72(12):928–931
51. Manohar S, Dahar K, Adler HJ, Dalian D, Salvi R (2016) Noise-induced hearing loss: neuropathic pain via Ntrk1 signaling. Mol Cell Neurosci 75:101–112
52. Khalfa S, Bruneau N, Roge B, Georgieff N, Veuillet E, Adrien JL et al (2004) Increased perception of loudness in autism. Hear Res 198(1–2):87–92
53. Gomes E, Pedroso FS, Wagner MB (2008) Auditory hypersensitivity in the autistic spectrum disorder. Pro Fono 20(4):279–284
54. Geisser ME, Glass JM, Rajcevska LD, Clauw DJ, Williams DA, Kileny PR et al (2008) A psychophysical study of auditory and pressure sensitivity in patients with fibromyalgia and healthy controls. J Pain 9(5):417–422
55. Gothelf D, Farber N, Raveh E, Apter A, Attias J (2006) Hyperacusis in Williams syndrome: characteristics and associated neuroaudiologic abnormalities. Neurology 66(3):390–395
56. Ashkenazi A, Yang I, Mushtaq A, Oshinsky ML (2010) Is phonophobia associated with cutaneous allodynia in migraine? J Neurol Neurosurg Psychiatry 81(11):1256–1260
57. Irimia P, Cittadini E, Paemeleire K, Cohen AS, Goadsby PJ (2008) Unilateral photophobia or phonophobia in migraine compared with trigeminal autonomic cephalalgias. Cephalalgia 28(6):626–630
58. Juris L, Andersson G, Larsen HC, Ekselius L (2013) Psychiatric comorbidity and personality traits in patients with hyperacusis. Int J Audiol 52(4):230–235
59. Baguley D, McFerran D, Hall D (2013) Tinnitus. Lancet 382(9904):1600–1607
60. Malouff JM, Schutte NS, Zucker LA (2011) Tinnitus-related distress: a review of recent findings. Curr Psychiatry Rep 13(1):31–36
61. Langguth B, Landgrebe M, Kleinjung T, Sand GP, Hajak G (2011) Tinnitus and depression. World J Biol Psychiatry 12(7):489–500
62. Loprinzi PD, Maskalick S, Brown K, Gilham B (2013) Association between depression and tinnitus in a nationally representative sample of US older adults. Aging Ment Health 17(6):714–717

63. Attanasio G, Russo FY, Roukos R, Covelli E, Cartocci G, Saponara M (2013) Sleep architecture variation in chronic tinnitus patients. Ear Hear 34(4):503–507
64. Alster J, Shemesh Z, Ornan M, Attias J (1993) Sleep disturbance associated with chronic tinnitus. Biol Psychiatry 34(1–2):84–90
65. Hallpike CS, Hood JD (1959) Observations upon the neurological mechanism of the loudness recruitment phenomenon. Acta Otolaryngol 50:472–486
66. Davis H, Morgan CT, Hawkins JE, Galambos R, Smith FW (1950) Temporary deafness following exposures to loud tones and noise. Acta Otolaryngol (Stockholm) Suppl 88:1–57
67. van der Staay FJ, Arndt SS, Nordquist RE (2009) Evaluation of animal models of neurobehavioral disorders. Behav Brain Funct 5:11
68. Festing MF (2004) Is the use of animals in biomedical research still necessary in 2002? Unfortunately, "yes". Altern Lab Anim 32(Suppl 1B):733–739
69. Weiler EM, Sandman DE, Pederson LM (1981) Magnitude estimates of loudness adaption at 60 dB SPL. Br J Audiol 15(3):201–204
70. Epstein M, Florentine M (2006) Loudness of brief tones measured by magnitude estimation and loudness matching. J Acoust Soc Am 119(4):1943–1945
71. Kamm C, Dirks DD, Mickey MR (1978) Effect of sensorineural hearing loss on loudness discomfort level and most comfortable loudness judgments. J Speech Hear Res 21(4):668–681
72. Hood JD (1977) Loudness balance procedures for the measurement of recruitment. Audiology 16(3):215–228
73. Miskolczy-Fodor F (1953) Monaural loudness balance-test and determination of recruitment-degree with short sound-impulses. Acta Otolaryngol 43(6):573–595
74. Schecklmann M, Lehner A, Schlee W, Vielsmeier V, Landgrebe M, Langguth B (2015) Validation of screening questions for hyperacusis in chronic tinnitus. Biomed Res Int 2015:191479
75. Khalfa S, Dubal S, Veuillet E, Perez-Diaz F, Jouvent R, Collet L (2002) Psychometric normalization of a hyperacusis questionnaire. ORL J Otorhinolaryngol Relat Spec 64(6):436–442
76. Wallen MB, Hasson D, Theorell T, Canlon B (2012) The correlation between the hyperacusis questionnaire and uncomfortable loudness levels is dependent on emotional exhaustion. Int J Audiol 51(10):722–729
77. Fackrell K, Fearnley C, Hoare DJ, Sereda M (2015) Hyperacusis questionnaire as a tool for measuring hypersensitivity to sound in a tinnitus research population. Biomed Res Int 2015:290425
78. Ida-Eto M, Hara N, Ohkawara T, Narita M (2017) Mechanism of auditory hypersensitivity in human autism using autism model rats. Pediatr Int 59(4):404–407
79. Zerbi V, Ielacqua GD, Markicevic M, Haberl MG, Ellisman MH, A AB et al (2018) Dysfunctional autism risk genes cause circuit-specific connectivity deficits with distinct developmental trajectories. Cereb Cortex 28(7):2495–2506
80. Liao W, Gandal MJ, Ehrlichman RS, Siegel SJ, Carlson GC (2012) MeCP2+/− mouse model of RTT reproduces auditory phenotypes associated with Rett syndrome and replicate select EEG endophenotypes of autism spectrum disorder. Neurobiol Dis 46(1):88–92
81. Lingenhohl K, Friauf E (1994) Giant neurons in the rat reticular formation: a sensorimotor interface in the elementary acoustic startle circuit? J Neurosci 14(3 Pt 1):1176–1194
82. Lee Y, Lopez DE, Meloni EG, Davis M (1996) A primary acoustic startle pathway: obligatory role of cochlear root neurons and the nucleus reticularis pontis caudalis. J Neurosci 16(11):3775–3789
83. Parham K, Willott JF (1990) Effects of inferior colliculus lesions on the acoustic startle response. Behav Neurosci 104(6):831–840
84. Leitner DS, Cohen ME (1985) Role of the inferior colliculus in the inhibition of acoustic startle in the rat. Physiol Behav 34(1):65–70
85. Rosen JB, Davis M (1988) Temporal characteristics of enhancement of startle by stimulation of the amygdala. Physiol Behav 44(1):117–123

86. Hormigo S, Lopez DE, Cardoso A, Zapata G, Sepulveda J, Castellano O (2018) Direct and indirect nigrofugal projections to the nucleus reticularis pontis caudalis mediate in the motor execution of the acoustic startle reflex. Brain Struct Funct 223(6):2733–2751

87. Fendt M, Koch M, Schnitzler HU (1994) Lesions of the central gray block the sensitization of the acoustic startle response in rats. Brain Res 661(1–2):163–173

88. Fendt M, Koch M, Schnitzler HU (1994) Sensorimotor gating deficit after lesions of the superior colliculus. Neuroreport 5(14):1725–1728

89. Blaszczyk JW, Tajchert K (1997) Effect of acoustic stimulus characteristics on the startle response in hooded rats. Acta Neurobiol Exp (Wars) 57(4):315–321

90. Jastreboff PJ, Brennan JF, Sasaki CT (1988) An animal model for tinnitus. Laryngoscope 98(3):280–286

91. Lobarinas E, Sun W, Cushing R, Salvi R (2004) A novel behavioral paradigm for assessing tinnitus using schedule-induced polydipsia avoidance conditioning (SIP-AC). Hear Res 190(1–2):109–114

92. Sun W, Lu J, Stolzberg D, Gray L, Deng A, Lobarinas E et al (2009) Salicylate increases the gain of the central auditory system. Neuroscience 159(1):325–334

93. Chen GD, Radziwon KE, Kashanian N, Manohar S, Salvi R (2014) Salicylate-induced auditory perceptual disorders and plastic changes in nonclassical auditory centers in rats. Neural Plast 2014:658741

94. Falbe-Hansen J (1941) Clinical and experimental histological studies of the effects of salicylates and quinine on the ear. Acta Otolaryngol 44(Suppl):1–216

95. Chen YC, Chen GD, Auerbach BD, Manohar S, Radziwon K, Salvi R (2017) Tinnitus and hyperacusis: contributions of paraflocculus, reticular formation and stress. Hear Res 349:208–222

96. Xiong B, Alkharabsheh A, Manohar S, Chen GD, Yu N, Zhao X et al (2017) Hyperexcitability of inferior colliculus and acoustic startle reflex with age-related hearing loss. Hear Res 350:32–42

97. Ison JR, Allen PD (2003) Low-frequency tone pips elicit exaggerated startle reflexes in C57BL/6J mice with hearing loss. J Assoc Res Otolaryngol 4(4):495–504

98. Dehmel S, Eisinger D, Shore SE (2012) Gap prepulse inhibition and auditory brainstem-evoked potentials as objective measures for tinnitus in guinea pigs. Front Syst Neurosci 6:42

99. Sun W, Deng A, Jayaram A, Gibson B (2012) Noise exposure enhances auditory cortex responses related to hyperacusis behavior. Brain Res 1485:108–116

100. Chen G, Lee C, Sandridge SA, Butler HM, Manzoor NF, Kaltenbach JA (2013) Behavioral evidence for possible simultaneous induction of hyperacusis and tinnitus following intense sound exposure. J Assoc Res Otolaryngol 14(3):413–424

101. Salloum RH, Yurosko C, Santiago L, Sandridge SA, Kaltenbach JA (2014) Induction of enhanced acoustic startle response by noise exposure: dependence on exposure conditions and testing parameters and possible relevance to hyperacusis. PLoS One 9(10):e111747

102. Lobarinas E, Hayes SH, Allman BL (2013) The gap-startle paradigm for tinnitus screening in animal models: limitations and optimization. Hear Res 295:150–160

103. Knudson IM, Melcher JR (2016) Elevated acoustic startle responses in humans: relationship to reduced loudness discomfort level, but not self-report of hyperacusis. J Assoc Res Otolaryngol 17(3):223–235

104. Filion PR, Margolis RH (1992) Comparison of clinical and real-life judgement of loudness discomfort. J Am Acad Audiol 3:193–199

105. Arieh Y, Marks LE (2003) Recalibrating the auditory system: a speed-accuracy analysis of intensity perception. J Exp Psychol Hum Percept Perform 29(3):523–536

106. Wagner E, Florentine M, Buus S, McCormack J (2004) Spectral loudness summation and simple reaction time. J Acoust Soc Am 116(3):1681–1686

107. Leibold LJ, Werner LA (2002) Relationship between intensity and reaction time in normal-hearing infants and adults. Ear Hear 23:92–97

108. Marshall L, Brandt JF (1980) The relationship between loudness and reaction time in normal hearing listeners. Acta Otolaryngol 90(3–4):244–249
109. Pfingst BE, Hienz R, Kimm J, Miller J (1975) Reaction-time procedure for measurement of hearing. I. Suprathreshold functions. J Acoust Soc Am 57(2):421–430
110. Moody DB (1973) Behavioral studies of noise-induced hearing loss in primates: loudness recruitment. Adv Otorhinolaryngol 20:82–101
111. May BJ, Little N, Saylor S (2009) Loudness perception in the domestic cat: reaction time estimates of equal loudness contours and recruitment effects. J Assoc Res Otolaryngol 10(2):295–308
112. Lauer AM, Dooling RJ (2007) Evidence of hyperacusis in canaries with permanent hereditary high-frequency hearing loss. Semin Hear 28(4):319–326
113. Stebbins WC (1966) Auditory reaction time and the derivation of equal loudness contours for the monkey. J Exp Anal Behav 9(2):135–142
114. Stebbins WC, Miller JM (1964) Reaction time as a function of stimulus intensity for the monkey. J Exp Anal Behav 7:309–312
115. Radziwon K, Holfoth D, Lindner J, Kaier-Green Z, Bowler R, Urban M et al (2017) Salicylate-induced hyperacusis in rats: dose- and frequency-dependent effects. Hear Res 350:133–138
116. Florentine M, Buus S, Robinson M (1998) Temporal integration of loudness under partial masking. J Acoust Soc Am 104(2 Pt 1):999–1007
117. Moody DB (1970) Reaction time as an index of sensory function in animals. In: Stebbins WC (ed) Animal pscyhophysics: the design and conduct of sensory experiments. Appleton-Century-Crofts, New York, pp 277–301
118. Hall JW 3rd, Grose JH (1997) The relation between gap detection, loudness, and loudness growth in noise-masked normal-hearing listeners. J Acoust Soc Am 101(2):1044–1049
119. Feitosa AG, Moody DB, Stebbins WC (1981) Loudness recruitment in the dihydrostreptomycin-treated patas monkey. J Acoust Soc Am 70:s27
120. Pugh JE Jr, Moody DB, Anderson DJ (1979) Electrocochleography and experimentally induced loudness recruitment. Arch Otorhinolaryngol 224(3–4):241–255
121. Danesh AA, Lang D, Kaf W, Andreassen WD, Scott J, Eshraghi AA (2015) Tinnitus and hyperacusis in autism spectrum disorders with emphasis on high functioning individuals diagnosed with Asperger's syndrome. Int J Pediatr Otorhinolaryngol 79(10):1683–1688
122. Myers EN, Bernstein JM (1965) Salicylate ototoxicity; a clinical and experimental study. Arch Otolaryngol 82(5):483–493
123. Radziwon KE, Stolzberg DJ, Urban ME, Bowler RA, Salvi RJ (2015) Salicylate-induced hearing loss and gap detection deficits in rats. Front Neurol 6:31
124. Zhang C, Flowers E, Li JX, Wang Q, Sun W (2014) Loudness perception affected by high doses of salicylate—a behavioral model of hyperacusis. Behav Brain Res 271:16–22
125. Sun W, Fu Q, Zhang C, Manohar S, Kumaraguru A, Li J (2014) Loudness perception affected by early age hearing loss. Hear Res 313:18–25
126. Alkharabsheh A, Xiong F, Xiong B, Manohar S, Chen G, Salvi R et al (2017) Early age noise exposure increases loudness perception—a novel animal model of hyperacusis. Hear Res 347:11–17
127. Hayes SH, Radziwon KE, Stolzberg DJ, Salvi RJ (2014) Behavioral models of tinnitus and hyperacusis in animals. Front Neurol 5:179
128. Formby C, Sherlock LP, Gold SL (2003) Adaptive plasticity of loudness induced by chronic attenuation and enhancement of the acoustic background. J Acoust Soc Am 114(1):55–58
129. Brotherton H, Plack CJ, Schaette R, Munro KJ (2017) Using acoustic reflex threshold, auditory brainstem response and loudness judgments to investigate changes in neural gain following acute unilateral deprivation in normal hearing adults. Hear Res 345:88–95
130. Ruttiger L, Ciuffani J, Zenner HP, Knipper M (2003) A behavioral paradigm to judge acute sodium salicylate-induced sound experience in rats: a new approach for an animal model on tinnitus. Hear Res 180(1–2):39–50

131. Mohrle D, Hofmeier B, Amend M, Wolpert S, Ni K, Bing D et al (2019) Enhanced central neural gain compensates acoustic trauma-induced cochlear impairment, but unlikely correlates with tinnitus and hyperacusis. Neuroscience 407:146–169
132. Blaesing L, Kroener-Herwig B (2012) Self-reported and behavioral sound avoidance in tinnitus and hyperacusis subjects, and association with anxiety ratings. Int J Audiol 51(8):611–617
133. Manohar S, Spoth J, Radziwon K, Auerbach BD, Salvi R (2017) Noise-induced hearing loss induces loudness intolerance in a rat active sound avoidance paradigm (ASAP). Hear Res 353:197–203
134. McFerran DJ, Baguley DM (2007) Acoustic shock. J Laryngol Otol 121(4):301–305
135. Van Campen LE, Dennis JM, Hanlin RC, King SB, Velderman AM (1999) One-year audiologic monitoring of individuals exposed to the 1995 Oklahoma City bombing. J Am Acad Audiol 10(5):231–247
136. de Klaver MJ, van Rijn MA, Marinus J, Soede W, de Laat JA, van Hilten JJ (2007) Hyperacusis in patients with complex regional pain syndrome related dystonia. J Neurol Neurosurg Psychiatry 78(12):1310–1313
137. Kusdra PM, Stechman-Neto J, Leao BLC, Martins PFA, Lacerda ABM, Zeigelboim BS (2018) Relationship between otological symptoms and TMD. Int Tinnitus J 22(1):30–34
138. Geisser ME, Strader Donnell C, Petzke F, Gracely RH, Clauw DJ, Williams DA (2008) Comorbid somatic symptoms and functional status in patients with fibromyalgia and chronic fatigue syndrome: sensory amplification as a common mechanism. Psychosomatics 49(3):235–242
139. Klune CB, Larkin AE, Leung VSY, Pang D (2019) Comparing the rat grimace scale and a composite behaviour score in rats. PLoS One 14(5):e0209467
140. Sneddon LU, Elwood RW, Adamo SA, Leach MC (2014) Defining and assessing animal pain. Anim Behav 97:201–212
141. Bacciottini L, Pellegrini-Giampietro DE, Bongianni F, de Luca G, Beni M, Politi V et al (1987) Biochemical and behavioural studies on indole-pyruvic acid: a keto-analogue of tryptophan. Pharmacol Res Commun 19(11):803–817
142. Hong R, Sur B, Yeom M, Lee B, Kim KS, Rodriguez JP et al (2018) Anti-inflammatory and anti-arthritic effects of the ethanolic extract of Aralia continentalis Kitag. in IL-1beta-stimulated human fibroblast-like synoviocytes and rodent models of polyarthritis and nociception. Phytomedicine 38:45–56
143. Helmstetter FJ, Bellgowan PS (1994) Hypoalgesia in response to sensitization during acute noise stress. Behav Neurosci 108(1):177–185
144. Stevens MW, Domer FR (1973) Alterations in morphine-induced analgesia in mice exposed to pain, light or sound. Arch Int Pharmacodyn Ther 206(1):66–75
145. Marone JG (1968) Suppression of pain by sound. Psychol Rep 22(3):1055–1056
146. Shankar N, Awasthy N, Mago H, Tandon OP (1999) Analgesic effect of environmental noise: a possible stress response in rats. Indian J Physiol Pharmacol 43(3):337–346
147. Benedek G, Szikszay M (1985) Sensitization or tolerance to morphine effects after repeated stresses. Prog Neuropsychopharmacol Biol Psychiatry 9(4):369–380
148. Gardner WJ, Licklider JC (1959) Auditory analgesia in dental operations. J Am Dent Assoc 59:1144–1149
149. Ramar K, Hariharavel VP, Sinnaduri G, Sambath G, Zohni F, Alagu PJ (2016) Effect of audioanalgesia in 6- to 12-year-old children during dental treatment procedure. J Contemp Dent Pract 17(12):1013–1015
150. Baghdadi ZD (2000) Evaluation of audio analgesia for restorative care in children treated using electronic dental anesthesia. J Clin Pediatr Dent 25(1):9–12
151. Blass BC (1975) Sound analgesia. J Am Podiatry Assoc 65(10):963–971
152. Handley LJA (1970) "Sound" approach to analgesia. J Rocky Mt Analg Soc 2(2):39–40
153. Horowitz LG (1992) Audiotaped relaxation, implosion, and rehearsal for the treatment of patients with dental phobia. Gen Dent 40(3):242–247

154. Simkin P, Bolding A (2004) Update on nonpharmacologic approaches to relieve labor pain and prevent suffering. J Midwifery Womens Health 49(6):489–504
155. Kittelberger JM, Bass AH (2013) Vocal-motor and auditory connectivity of the midbrain periaqueductal gray in a teleost fish. J Comp Neurol 521(4):791–812
156. Algom D, Raphaeli N, Cohen-Raz L (1986) Integration of noxious stimulation across separate somatosensory communications systems: a functional theory of pain. J Exp Psychol Hum Percept Perform 12(1):92–102
157. Job A, Pons Y, Lamalle L, Jaillard A, Buck K, Segebarth C et al (2012) Abnormal cortical sensorimotor activity during "Target" sound detection in subjects with acute acoustic trauma sequelae: an fMRI study. Brain Behav 2(2):187–199
158. Job A, Paucod JC, O'Beirne GA, Delon-Martin C (2011) Cortical representation of tympanic membrane movements due to pressure variation: an fMRI study. Hum Brain Mapp 32(5):744–749
159. Kessels RP, Keyser A, Verhagen WI, van Luijtelaar EL (1998) The whiplash syndrome: a psychophysiological and neuropsychological study towards attention. Acta Neurol Scand 97(3):188–193
160. Tanaka N, Atesok K, Nakanishi K, Kamei N, Nakamae T, Kotaka S et al (2018) Pathology and treatment of traumatic cervical spine syndrome: whiplash injury. Adv Orthop 2018:4765050
161. Sterling M, Jull G, Vicenzino B, Kenardy J (2003) Sensory hypersensitivity occurs soon after whiplash injury and is associated with poor recovery. Pain 104(3):509–517
162. Chen GD, Manohar S, Salvi R (2012) Amygdala hyperactivity and tonotopic shift after salicylate exposure. Brain Res 1485:63–76
163. Auerbach BD, Radziwon K, Salvi R (2019) Testing the central gain model: loudness growth correlates with central auditory gain enhancement in a rodent model of hyperacusis. Neuroscience 407:93–107
164. Levitin DJ, Menon V, Schmitt JE, Eliez S, White CD, Glover GH et al (2003) Neural correlates of auditory perception in Williams syndrome: an fMRI study. Neuroimage 18(1):74–82
165. Asokan MM, Williamson RS, Hancock KE, Polley DB (2018) Sensory overamplification in layer 5 auditory corticofugal projection neurons following cochlear nerve synaptic damage. Nat Commun 9(1):2468

New Automatic and Robust Measures to Evaluate Hearing Loss and Tinnitus in Preclinical Models

A. Laboulais, S. Malmström, C. Dejean, M. Cardoso, T. Le Meur, L. Almeida, C. Goze-Bac, and S. Pucheu

Abbreviations

BBN	Broadband noise
BOLD	Blood oxygen level dependent
CC	Cerebral cortex
FDA	US Food and Drug Administration
fMRI	Functional magnetic resonance imaging
GPIAS	GAP inhibition of the acoustic startle reflex
HCs	Hair cells
IC	Inferior colliculus
IHC	Inner hair cells
IP	Intraperitoneal
MEMRI	Manganese enhancement magnetic resonance magnetic
$MnCl_2$	Manganese chloride
NMR	Nuclear magnetic resonance
OHC	Outer hair cells
PET	Positron emission tomography
$R2$	NMR relaxation rate
rCBF	Regional cerebral blood flow

A. Laboulais
Charles Coulomb Laboratory (L2C-BioNanoNMRI Team), UMR 5221 Centre National de la Recherche Scientifique—University, Montpellier, France

CILcare, Advanced Solution for Drug Development in Hearing Disorder, Montpellier, France

S. Malmström · C. Dejean · S. Pucheu (✉)
CILcare, Advanced Solution for Drug Development in Hearing Disorder, Montpellier, France
e-mail: sylvie.pucheu@cilcare.com

M. Cardoso · C. Goze-Bac
Charles Coulomb Laboratory (L2C-BioNanoNMRI Team), UMR 5221 Centre National de la Recherche Scientifique—University, Montpellier, France

T. Le Meur · L. Almeida
Keeneye, Paris, France

© Springer Nature Switzerland AG 2020
S. Pucheu et al. (eds.), *New Therapies to Prevent or Cure Auditory Disorders*,
https://doi.org/10.1007/978-3-030-40413-0_7

RGB   Red green blue
r_i Relaxivity
ROI   Region of interest
SCs   Supporting cells
SEM   Standard error of the mean
SGNs   Spiral ganglion neurons
SIR   Signal intensity ratio
SIT   Salicylate-induced tinnitus
SNR   Signal-to-noise ratio
$T2$ Transversal relaxation time
TT   Transtympanic

1   Introduction

It is estimated that more than 500 million persons worldwide suffer from hearing disorders, which can be due to many different factors such as acoustic trauma, aging, or drug toxicity [1, 2]. Hearing loss, the most common sensory handicap, occurs when structures in the inner ear (the stria vascularis, outer hair cells (OHC), or inner hair cells (IHC) in the organ of Corti or spiral ganglion neurons (SGNs)) are damaged [1]. The most widely used techniques for assessing preclinical hearing and hearing impairments are functional electrophysiological (evoked potential) and electroacoustic (distortion products) measurements, or morphological observations that involve histological examinations or cell counting from tissues in the cochlea or brain. One of the most common histological measures is the cochleogram, which consists of evaluating and counting hair cells (HC). The first difficulty of the cochleogram technique lies in obtaining a preparation of good quality, ideally one that encompasses the entire length of the basilar membrane from the high-frequency base to the low-frequency apex. The second difficulty is making accurate counts of HC, both those that are considered present and those that are absent. The counting procedure is typically performed manually or in some cases with a semiautomated method. The manual counting procedure requires much attention and expertise, is time-consuming, and potentially subject to observer bias. These limitations make manual, observer-based counts problematic for use in preclinical research and the development of new therapeutic solutions.

Tinnitus, another major hearing disorder that can be disabling, is a phantom auditory sensation which occurs in the absence of an external sound stimulus. Currently, one-third of the population experiences tinnitus at least once in their life and 1–5% develop persistent tinnitus with serious complications [3]. One major obstacle for detecting tinnitus in animal models arises from the fact that tinnitus is a subjective phenomenon. In humans, subjective tinnitus characteristics and severity can be obtained through questionnaires and self-reports [4]. In animal models, researchers have developed various behavioral techniques to evaluate tinnitus [5].

However, many of these behavioral models are time-consuming because they require a lot of animal training and seldom give quantitative results related to the loudness, pitch, or severity of the tinnitus percept. To circumvent these difficulties, the scientific community is developing in parallel neuroimaging techniques to observe neural signatures of auditory activities. The first utilization of the manganese enhanced magnetic resonance imaging (MEMRI) for the detection of tinnitus in animal models was reported in 2005 [6–8]. Nevertheless, more robust and standardized MEMRI acquisition protocols and analysis methods need to be established. Indeed, there are no standard tools available for automated procedures in tinnitus analyses. The following chapter presents recent developments in these research areas, aiming to develop new automatic techniques that are efficient and well-controlled to produce cochleogram to detect hearing loss and robust procedures to quantify by MEMRI the presence of tinnitus in animal models.

2   Evaluation of Human and Computer-Assisted Quantification of Hair Cells in Full-Length Cochlea

2.1   Introduction

The relationship between cochlear HC functional readouts and actual HC damage or loss after cochlear injury is complex and not straightforward [9–11]. Nevertheless, in preclinical research the cochleogram is fundamental tool used to quantify HC damage, evaluate treatments aimed at limiting the loss of HC or testing methods for inducing their HC regeneration. Moreover, in pharmacological safety and the evaluation of ototoxicity of new or repositioned drugs, the cochleogram is one of the assays required by the FDA [12].

The cochleogram technique has been widely used for decades by academic researchers, but with no standardization of techniques or counting procedures [13]. These authors suggested a standardization of the cochleogram procedure stating that: (1) basilar membrane length should be plotted as a percentage instead of in millimeters, due to the biological variations that exist in basilar membrane length within a particular species and strain, and the total length in millimeters stated on the cochleogram and (2) the equations used for frequency–place maps should be stated on the cochleogram.

One of the difficulties of the cochleogram technique lies in obtaining a good quality preparation of the sensory epithelium containing the organ of Corti and its rows of HC. In humans, the cochlea is located in the temporal bone, the hardest bone in the body. Two major categories of tissue preparation are commonly used to evaluate the sensory epithelium, either embedding and sectioning or careful dissection of the sensory epithelium as a flat surface (or whole mount) preparations [14]. In both methods, the HCs can be labeled with different staining protocols, histological stains, fluorescent markers or immunolabeling methods, and visualized using a variety of microscopic techniques. The preparation consists of embedding the cochlea in a medium, usually paraffin or plastic, and then serially sectioning it to expose the hair cells [14–16]. The sections can be either parallel or perpendicular to the modiolar axis.

For the flat surface preparation method, the organ of Corti is extracted from the cochlea by microdissection into cochlear turns or half-turns. The cochlear fragments are prepared as flatly as possible by trimming off surrounding structures and mounting the specimens on a microscope slide. The labeled hair cells are then visualized for counting along the entire length of the organ of Corti [13, 14]. For the assessment of HC survival and construction of a cochleogram, the flat surface method is often preferred, due to its relative simplicity, comprehensive assessment over the length of the cochlea and the speed of cell counting; moreover, immunohistochemistry can be more easily performed on dissected tissue than on plastic embedded sections [14].

The HC counting procedure is typically performed manually or with a semiautomated method. In mouse cochleae, there are ~700 IHCs and ~2400 OHCs [17, 18] and in rat cochleae from the Sprague–Dawley strain there are ~1000 IHCs and 3700 OHCs [19]. In the guinea pig cochleae, there are approximately three times more HCs than in the mouse and two times more than in the rat (~2000 IHCs and ~7000 OHCs) [20]. The manual counting procedure requires much attention and expertise and is very time-consuming. The loss of cochlear hair cells has been widely used in various experimental animal models, including guinea pigs, chinchillas, rats, and mice [21–25].

In the field of preclinical development of new therapies, there is great interest to develop new automatic techniques for counting hair cells that are rapid, efficient, unbiased, and well controlled. In this context, we proceeded to develop an algorithm capable of automatically classifying and counting HCs in cochlear fragment images from flat surface preparations. To validate and determine its accuracy, the automatic results were compared to the results from manual counting performed by several experts. Many characteristics inherent to the input data such as the type of image, the morphologic characteristics of the objects, or unknown user accuracy are required to develop a specific pipeline for this project.

2.2   Material and Methods

Generation of Data

Cochleae were sampled from three control and three cisplatin-treated Wistar rats (10 mg/kg, single IP injection). Three days after cisplatin treatment, the cochleae were removed and immediately fixed in 4% PFA and then decalcified in 10% EDTA for several days. The membranous and sensory spiral containing the organ of Corti was dissected using a flat surface preparation yielding 5–6 fragments per cochlea. The hair cells in each fragment were immunolabeled using a primary antibody against Myosin-VIIa, and a fluorescent secondary antibody (protocol adapted from Akil and Lustig [21]). The fragments were then mounted on a microscope slide and images were acquired on a confocal microscope (20× objective). Each x–y image plane contained approximately 2000 × 3000 pixels, with a two-dimensional resolution of 0.642 µm per pixel for each focal plane. Approximately 30 focal planes were obtained per fragment, each plane separated by 1.64 µm.

Three independent scientists, highly experienced in cochlear cell counting, independently manually counted and annotated all IHCs and OHCs in each cochlear fragment in six cochleae (three control and three cisplatin). Users were instructed to annotate every labeled cell with a point annotation within the cell's three-dimensional center and identify them either as an IHC or OHC. For the OHCs, their respective row location (1, 2 or 3) was taken into account. Users were blinded to the treatments. Fragments were identified by a unique number that specified both the cochlea and fragment number. The labeled point annotation was obtained for each cochlear fragment, with the exception of fragments in which the user could not identify any cells as being present. Majority voting (two out of three) based on the value for each cell was used to determine a unique set of point annotations for each fragment.

Algorithm Development

Due to the volumetric nature of the image data, together with the fact that users annotated the cells using a 2D stack viewer, we developed a 3D object detection algorithm that would perform detections on each stack. A model based on a deep convolutional neural network (CNN) was first used in order to get the detections for each stack of 3D images [26]. The last feature map of the CNN allowed us to predict the cell's locations and class (IHC or OHC). Predictions were then grouped using an unsupervised method to merge two-dimensional predictions into a subset of unique 3D hair cell detection specification.

The image's characteristics directly impact the development of the algorithm. The difference between the pixel size along the depth (between each stack) and the pixel size on the 2D x–y plan must be carefully taken into account in order to provide consistent detections. The whole 3D images cannot be loaded and analyzed at once by the network for both training and routine use due to their excessive sizes. Preprocessing the images before using them in the algorithm is therefore a necessity. In addition, the image quality can vary depending on several parameters such as the treatment of the animals or if they originate from healthy controls or cochleae damaged by noise or ototoxic agents. The fragment location within the cochlea can also affect the quality of the final image. The first fragments from apical or middle regions of the cochlea tend to be less damaged than fragments from the basal regions, where damage due to dissection, preparation, or treatment is more likely to occur. In these cases, the context information for the cell detection is degraded or completely lost due to extensive damage.

As previously discussed, the characteristics of the objects are contained in their relative locations in the three-dimensional volume and their morphology is revealed by the staining procedures. All hair HCs are small relative to the size of the image and share large similarities between the two classes (i.e., IHCs vs. OHCs). In order to perform proper classification, the algorithm needs to be aware of a large context area surrounding each object, because the context area provides important criteria for differentiating the different cell types. Moreover, OHCs and IHCs are not equally distributed within the dataset because there are approximately three times more OHC than IHC per fragment image. The unbalanced dataset makes the learning process of the network more difficult for the underrepresented objects.

In order to train the 3D deep convolutional object detection network, we used five different fragments from two separate cochleae, one control and one cisplatin treated. Each fragment is comprised of a stack of 35 slices of various dimensions (approximately 3000×2000 pixels). To limit bandwidth during training, each 3D image was subdivided into smaller images. However, the subregion was sufficiently large that it included both IHCs and OHCs. These subregions covered an area of $320\ \mu m \times 320\ \mu m$. Each expert only annotated the cell's center with a point in order to reduce the annotation time and have a sufficient amount of data for the training process. Using the average size of the cells, the point annotations are added to 3D box annotations. Those last annotations are then used by the network as localization ground truth (GT) information, i.e., the position and size that should be inferred. The overall dataset was partitioned in two sub datasets. Two cochleae containing 90% of the subregions were used for training and cross-validation, while 10% was set aside for evaluation of statistical metrics. Then, four separate cochleae were used for full cochleogram comparisons. To increase the number of examples without increasing the amount of initial data, image augmentations such as rotations, scaling, and contrast manipulations were performed on the training set in order to further train the model.

Transfer learning is a widely used technique in deep learning, which was used to provide a basic knowledge to the network before training to save time and data [27]. The convolutional neural network is initialized with another network pretrained on a similar task. During the training session, the model parameters are optimized to reduce the error of cell localization and classification with regard to the ground truth annotations. This optimization is done with a gradient descent algorithm called stochastic gradient descent (SGD) with momentum [28]. While accumulating error information from the previous predictions, it enables training with only one 3D image at a time without instability.

In order to assess the performance of the algorithm, a comparison was performed between manual and automated counting of hair cells. Two cochleograms with IHC and OHC were computed from the two sets of annotations, one evaluated by humans and one evaluated algorithmically using a line drawn by human experts along the IHCs of each cochlear fragment. The line length for each fragment was converted into a percentage regarding the total cochlear length and the density of cells was computed for each 5% interval. The average and standard deviation (SD) for hair cell density were then calculated and plotted for each percent distance.

2.3   Results

Algorithm Development Results

The quality of the training was evaluated with a precision–sensitivity curve for each class. The precision–sensitivity curve, also called precision–recall curve, summarizes the trade-offs between the true positive rate (sensitivity) and the precision of the model using different probability thresholds. For each probability, the number of true positive detections is defined using a specific matching method. A distance threshold vector, representing the average semiaxis values of a cell along each dimension is used to define whether or not a detection matches a ground truth annotation. Figure 1 shows the precision–sensitivity curves for IHC and OHC objects. The IHC curve shows a drop of precision for low sensitivity due to cells correctly detected but counted as false positive because it was not annotated by the expert. The two optimal points of these curves (red dots in Fig. 1) were then used to get the optimal probability threshold for each class. To provide quality results when running the algorithm on unknown fragments, cell detections were filtered with those thresholds regarding their confidence probability outputted by the network.

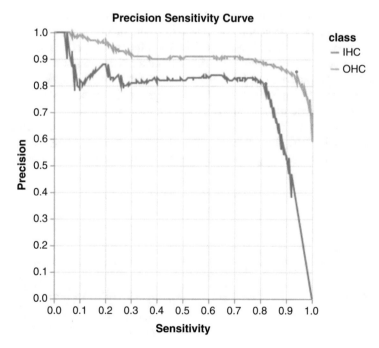

Fig. 1 Precision–sensitivity curves for IHC and OHC objects on the test subregions. Note the lower overall performance for IHCs is due to the bias of over sampling of OHCs

Comparison of Counting Time

One of the main advantages of an algorithmic approach is speed, reliability, and reproducibility. Regarding time, each user took on average 30.6 ± 6.5 min to fully annotate a fragment containing on average 887 ± 125 cells (Fig. 2). The algorithm, run through the KeenEye Platform, spent 3.5 ± 1.2 min to finalize the same analysis, showing a ~10-fold gain of time. This time represents the incompressible time for the algorithm to detect the hair cells, but also the data preparation and the post processing done in order to display the detection on the platform. For the automatic analysis, the small variation of duration on the fragment images is highly related to the size of the image; the larger the image, the longer the counting time.

Quality of Hair Cell Counting

The difficulty of the counting task can be illustrated by the variability of the IHC and OHC counts among the different experts. The count variability among the experts, measured as the standard deviation of counts over the mean of counts per fragment, was (median ± SD) 8.9 ± 13% for IHCs and 18.3 ± 33% for OHCs counts (Fig. 3).

The algorithm performance measures were quantitatively measured on the test set with two metrics: precision and sensitivity. The precision represents the percentage of correct results, while the sensitivity refers to the percentage of ground truth object correctly detected. The fully automated detection and counting software achieved a precision of 89 ± 9% and sensitivity of 69 ± 12%; the relative distributions are shown in the box–whisker diagrams in Fig. 4. The high precision of the algorithm ensures that each one of its detections is close to a ground truth annotation; however, this lowers the sensitivity and increases the variability. This result is explained by the difficulties the algorithm has performing correct cell detection on damaged fragments where the context information is missing or corrupted.

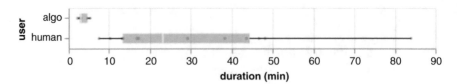

Fig. 2 Time in minutes for annotating each fragment among human users compared to the algorithm

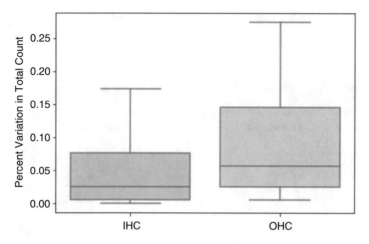

Fig. 3 Box–Whisker (median ± SD) distributions for % variability of counts for manual counting

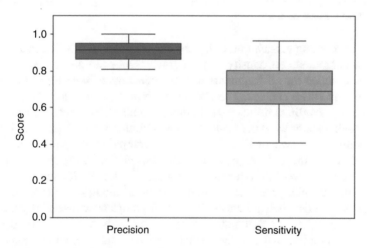

Fig. 4 Box–Whisker plots (median ± SD) for the sensitivity and precision scores for the algorithm versus test cases

Cochleogram Analysis

The cochleogram analysis was performed on cisplatin-treated and control cochleae. Using a line drawn along the IHCs by an expert, we computed the full cochleograms resulting from the manual and automated counting and compared them to each other (Fig. 5). For this dataset, manual and automated counting procedures had similar standard deviations. Furthermore, the differences among cochleograms were not significant, with significance determined by a Mann–Whitney test comparing the distribution of differences between the human and

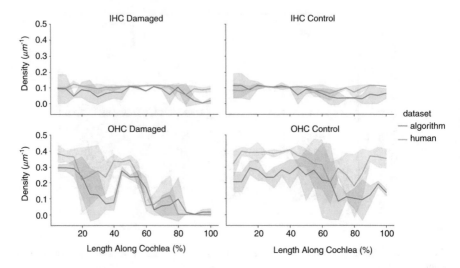

Fig. 5 Example comparing cochleograms obtained from human manual (orange) counting and automated (blue) counting of OHCs and ICHs in cisplatin-treated rats. Solid line is the mean with bands for the standard deviation

automatic counting at each bin. Using the closest point on this line to any detection, we computed cell density cochleograms. Once the cell count along this line was grouped for all fragments in a cochlea, counts were binned along the length of the entire cochlea. The size of the bins was chosen to be 5% of the total cochlea length, and thus varied among cochleae. The average and SD for the hair cell density were then calculated and plotted for each percent distance. The results are shown in Fig. 5, where we compared all expert annotations (denoted as human) to the algorithm results; this allowed us to see the variation in counts due to expert disagreement along the cochlea. Both counting methods resulted in a cochleogram with similar profiles all along the cochlea for IHCs and OHCs. The IHC density curves in the two groups were essentially the same with both manual and automated counting methods and close to 0.1 cell/ μM. Concerning the OHCs of the cisplatin-treated cochleae, a lower sensitivity was noted in the apical half of the cochlea with automated counting, while in the basal half both methods gave similar curves. For the control group, a difference in OHC density between the manual and automated counting procedures was observed all along the cochlea.

2.4 Discussion

Many studies rely on cochleogram analyses, but most of them use manual cell counting. A common procedure is to manually count the present and absent hair cells directly under the microscope using an eyepiece to measure and segment the basilar membrane. Hair cell loss is then plotted as a function of the distance from apex, either in mm or in length converted to % [13, 29]. A stereological approach has also been used on phalloidin-stained stereocilia bundles as counting units [30].

During this collaboration between CILcare and KeenEye Technologies, a full pipeline has been designed to automatically classify and quantify the number of hair cells in 3D cochlea confocal images based on flat surface preparation and fluorescent labeling of hair cells. This project introduces many challenges with regard to specific pre- and post-data processing and an adaptive model for 3D object detection. The model has been trained using transfer learning with mini batch images keeping the context information around the different types of cells. This new automatic counting method was 10 times faster than the human method, taking on average 3.5 min to analyze one fragment image. Indeed, even a trained expert performing manual counting of all labeled hair cells in a cochlea requires a minimal time to register the cell count, with a mouse click or a manual cell counter. Alternative methods exist to produce a cochleogram where only the missing cells and the length of the fragments are counted and the normal (ground truth) cell density is used to get the total number of cells. This method is relatively efficient for cochleae with little cell loss or a high degree of cell loss, but for intermediate cell loss, the counting procedure is longer.

It is important to note that due to the possible non-specificity of the staining, together with the fact that a single stain was used, cell identity is often determined using subtle information associated with the cells relative to their positions and morphology. For example, OHCs are grouped in three parallel rows, while the IHCs form a single row. Occasionally, extra OHCs may be observed radial to the third OHC row anywhere along the length of the organ of Corti. The organization of the OHCs is generally less ordered at the apical end than at the base or in the middle regions and is also dependent on possible organ health and/or damage. Cell identification becomes extremely subjective when the cochleae are highly damaged, or staining is incomplete, making unambiguous identification almost impossible. This ambiguity is explicitly shown in Fig. 6, where only a single IHC row has been annotated by only one expert.

The algorithm gave performance metrics of 90% for precision and 70% for sensitivity. While the precision is good, improvements are required to increase the overall sensitivity while reducing the variance. Developing an end-to-end 3D convolutional neural network to capture the 3D features of each cell could improve the performance of the algorithm making it more robust for analyzing images of different quality. However, this technique requires more computational memory resources and is unable to use large context information for cell classification. In order to improve detection results, object balancing could be added to the data preprocessing step before the training paradigm. Methods such as weighting the error for each detection regarding its ground truth class or underrepresented object augmentation can be used to solve this problem. However, the impact of these methods on the overall metrics has not been

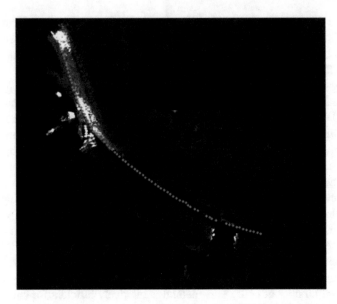

Fig. 6 Image of cochlear fragment where only a single row of IHCs (red dots) is annotated

established. The algorithm already performs properly on IHC detection due to greater context information awareness where the IHCs are anatomically more isolated and thus their identification is less ambiguous. At this stage, algorithm performance is highly correlated with the state of the fragments in the images. Fragments can be damaged during dissection, leading to images where cell integrity is not well preserved. The model was trained on images of good quality; however, the model performance decreases on damaged fragments as the context information is corrupted. Detection results on damaged images, and on the overall set of cochlear images, could be improved by adding damaged samples to the training dataset.

This algorithm, and the full pipeline associated with it, can be reused for future projects. The model can conserve its performances on images with small variations of characteristics regarding the initial fragment images it was trained on. This would be true for hair cell counting projects with similar cochlear images but coming from different species. However, if the image acquisition changes, it might cause important changes in the characteristics of the input images (e.g., number of channels, type of labeling, etc.). In this case, this algorithm can still be used via transfer learning, providing a knowledge base for a future one. The new training will be faster and will require a smaller amount of data to reach similar performances.

3   Comparison of Three Methods to Analyze Preclinical Manganese-Enhanced Magnetic Resonance Imaging on Salicylate-Induced Tinnitus in Rat Models

3.1   Introduction

The phantom sound of tinnitus is believed to be generated at certain points in the auditory system; it is generally triggered/initiated by cochlea damage, resulting in neural changes perceived as a phantom auditory sensation in the auditory cortex or other parts of the central nervous system. It can be triggered by noise-induced hearing loss, presbycusis, otosclerosis, otitis, Meniere's disease, or by ototoxic medications [31]. Tinnitus models in animals are mainly based on salicylate-induced tinnitus and noise-induced tinnitus. One major obstacle for the detection of tinnitus in these models arises from the fact that tinnitus is a subjective phenomenon, the only possible diagnosis relies on self-reports of the subjects in humans [4]. In animals, researchers have developed behavioral models using basic mechanisms of conditioning to evaluate tinnitus. It was first demonstrated by Jastreboff and colleagues [32] and later several extensions of that work have been published [5, 33]. Over the last few years, the GAP inhibition of the acoustic startle reflex (GPIAS) [34] has increasingly been used. GPIAS uses the amplitude of the acoustic startle reflex to a noise burst as a behavioral readout. The startle-evoking noise burst is presented on a background of moderate-intensity continuous noise. A silent interval, or gap, which serves as a pre-pulse, is presented shortly before the high-intensity noise burst. If a normal hearing animal hears the silent gap preceding the high-intensity noise burst, the startle reflex amplitude is inhibited. However, if an animal has tinnitus, the salience of the silent interval is reduced presumably because the phantom sound of tinnitus "fills in" the gap. In cases where the gap stimulus fails to significantly inhibit the amplitude of the startle reflex, the animal is presumed to have tinnitus. The frequency, bandwidth, and intensity of the carrier signal in which the gap is embedded can be varied with a goal of estimating the pitch and loudness of the tinnitus [35]. The GPIAS method does not require animal training. However, the methods used to collect and analyze GPIAS data vary between laboratories, making the comparison of results and reproducibility of data difficult [36]. In parallel, the auditory scientific community has become increasingly interested in developing imaging tools to detect tinnitus in animals. Indeed, neuroimaging techniques enable one to noninvasively visualize and study in vivo brain activities. Nowadays, the neuroimaging methods used in animal models include such techniques as manganese enhancement magnetic resonance imaging (MEMRI), functional magnetic resonance imaging (fMRI), and micro positron emission tomography (microPET). MEMRI and fMRI are based on nuclear magnetic resonance in strong magnetic fields to create various types of anatomical or functional brain images. MEMRI utilizes manganese chloride ($MnCl_2$) as a contrast agent to follow neuronal brain activity [37, 38]. Mn^{2+} ions enter active cells through voltage-gated calcium channels and accumulate inside neurons. Blood oxygen level dependent (BOLD) fMRI detects changes in deoxyhemoglobin concentrations under task-induced or spontaneous resting state conditions that reflect the degree of neural activity. PET imaging is based on the detection of radioactivity emitted by a tracer administered to the subject. With the appropriate tracer injected into blood circulation, it is possible to assess regional cerebral blood flow (rCBF) in the resting state or under task-based conditions.

These imaging methods can be compared in terms of advantages and drawbacks. In small animal imaging studies, high spatial resolution is essential because of the small size of the brain in rodents compared to humans. MEMRI can provide functional information with a resolution of about 100 μm, whereas PET resolution is limited by the size of the gamma ray detectors, providing millimeter resolution. fMRI has a higher spatial resolution than PET, but lower resolution compared to MEMRI (fMRI spatial resolution ~500 μm; MEMRI spatial resolution ~100 μm with high magnetic field strength). In addition, the contrast in MEMRI is more directly related to neural activity, whereas fMRI relies on an indirect neurovascular coupling to obtain neuronal information. In addition, the anesthesia used to avoid animal movement in fMRI can affect the neuronal responses, including activity in the auditory system [39–42]. In contrast, $MnCl_2$ is administered prior to imaging when the animals are awake, thereby avoiding the confounding effects of anesthesia. The utilization of MEMRI has been used since 2005 when it was discovered that the blood–brain barrier (BBB) need not be weakened in order to stimulate injected manganese to enter the brain and taken up into spontaneously active or stimulus-activated neurons [8]. Therefore, major auditory studies, including tinnitus, were carried out by using MEMRI in the past few years [6, 7, 35, 43, 44]. However, some tinnitus studies were performed with fMRI and PET on animal models [9, 45, 46].

Currently there are controversies in the literature regarding the pathological basis of tinnitus, and the most appropriate analytical method to identify the neural signature of tinnitus. To accelerate the development of new therapies, it is essential to establish more robust and standardized MEMRI analysis methods. Here we present an innovative quantification technique based on magnetic resonance properties, which relies on the transverse relaxation rates, namely $R2$, which depends on the constituents of the tissues and the presence of contrast agent. The goal of this method is to precisely detect the accumulation of paramagnetic Mn^{2+} ions accumulated in active neurons through voltage-gated calcium channels [47]. The direct measure of the NMR signal in specific regions of the brain allows to determine the percentage change of $R2$ between auditory areas of interest versus nonauditory regions. A positive $\Delta R2/R2$ indicates an increase of the nuclear magnetic resonance relaxation rate related to accumulation of Mn^{2+} ions in active neurons. In the following paragraph, the $\Delta R2/R2$ quantification is compared with some other available MEMRI analysis methods, namely the signal-to-noise ratio (SNR) and the signal intensity ratio (SIR) [7, 35, 44, 48, 49]. In this study, our analyses focused on a specific auditory structure, the inferior colliculus (IC). The IC plays important roles in processing auditory information received from both ears and then relaying the acoustic information to higher and lower brain regions [50]. The IC was also selected since several MEMRI studies reported an increase of neuronal activity in the inferior colliculus in animal models of tinnitus [7, 43, 48, 51]. The underlying hypothesis is that abnormal neuronal activity in the IC, as reflected in MEMRI, can provide a biomarker for the detection of tinnitus [6].

3.2 Material and Methods

Animals (Fig. 7)

Forty male Long Evans rats (8 weeks old, weighting 250 g) were maintained under specific pathogen-free conditions. Before starting the study, the rats were allowed to acclimate for 7 days in a colony maintained at 22 ± 6 °C and a standard light/dark cycle. The animals were used in three experimental protocols: The first series of animals (A) ($n = 10$, 2 groups) was injected with $MnCl_2$ (0.2 mmol/kg) via

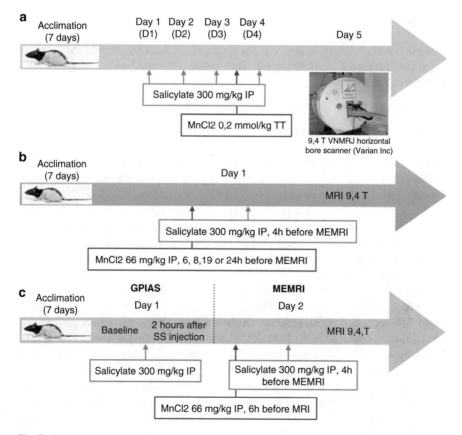

Fig. 7 Schematics of three different experimental protocols used in the study with $MnCl_2$ in which salicylate was administered by (**A**) transtympanic administration (**A**) and by (**B, C**) intraperitoneal route

transtympanic administration [52] 24 h before MEMRI measurement at fourth day (D4); the salicylate-induced tinnitus (SIT) group received sodium salicylate (300 mg/kg, i.p.) once a day from the first day (D1) to the fourth (D4), and a similar volume of saline was injected into sham group once a day from first day (D1) to fourth (D4) (Fig. 7) [3]. The second series (B) ($n = 24$, 8 groups) was injected with $MnCl_2$ (66 mg/kg) by an intraperitoneal route (i.p.) at 6, 8, 19, and 24 h before MEMRI measures. The SIT animals were treated with a single injection of sodium salicylate (300 mg/kg, i.p.) 4 h before MEMRI measurement. The sham group was injected with a similar volume of saline 4 h before MEMRI measurement. In the third series (C) ($n = 6$, 2 groups), GPIAS measurements were carried out in all animals, 2 h after administration of salicylate (300 mg/kg, i.p.) and before the MEMRI protocol ($MnCl_2$, i.p. 6 h, single injection of sodium salicylate, 300 mg/kg, i.p., 4 h before MEMRI measures) [3]. All animal experiments were conducted in accordance with European Directives (#2010/63/UE).

Salicylate

Salicylate was dissolved in sterile physiological saline at 176.5 mg/mL and administered at 300 mg/kg (i.p.). Sham salicylate animals received an injection of physiological saline.

MEMRI Acquisition Imaging

Before MRI image acquisition, rats were anesthetized using 2% isoflurane and monitored (respiration and temperature) using the MR-compatible Small Animal Monitoring and Gating System (Model 1025, SA Instruments, Inc., New York, USA). Respiration was maintained around 40 breaths/min by adjusting isoflurane level and oxygen flow rate. During the MRI acquisitions, the animal is maintained in a stable position with the use of a heated cradle equipped with head and tooth holder. The rat head is positioned at the center of the MRI magnet in a dedicated antenna fixed to the cradle.

MRI measurements were performed on a 400-MHz MRI scanner (Agilent Varian 9.4/160/ASR, Santa Clara, CA, USA) equipped with a MAGNEX TS1276D, a brain SHS coil 400 MHz (BioNanoNMRI, France) associated with an imaging acquisition system (Agilent, Palo Alto, California, USA), anesthetic and animal holder systems from Minerve Siemens A.G., Erlangen, Germany/RS2D (Haguenau, France). Axial images were acquired using a $T1$-weighted MEMS (Multiple Echo Multi Slices) protocol with the following parameters: Field of view, 60 × 30 mm matrix size, 512 × 256 axial slices with one zero filling, 1 mm slice thickness, no gap, TR = 500 ms, TE = 12.63 ms, number of averages = 8, number of slices = 16, scan time = 17 m and 45 s, NE = 2.

Gap Inhibition of Acoustic Startle Reflex

Behavioral testing was conducted using Kinder Scientific startle reflex hardware and software customized for this application by the manufacturer. Gap detection testing with background sounds presented through one speaker and startle stimuli presented through a second speaker located in the ceiling of the testing chamber 15 cm above the animal's head. The floor of the chamber is attached to a force transducer to measure the startle force applied to the floor. A clear polycarbonate animal holder, with holes cut for sound passage, was suspended above the floor allowing the animal to freely turn around while minimizing excessive movement. An adjustable-height roof was set to a level that keeps animals from rearing up, a behavior that adds variability to the startle response. Background sounds in the startle chamber consisted of 60 dB average SPL broadband noise (BBN) or bandpass-filtered noise. Each test session began with a 3-min acclimation period followed by two trials consisting of a startle-eliciting noise burst (115-dB SPL, 50-ms duration); the acclimation period served to habituate the startle response to a more stable baseline. Data from the first initial trial were not used in the gap detection analysis. The remainder of the session consisted of startle-only trials presented before or after gap trials in a counterbalanced manner throughout the session. Gap trials were identical to startle-only trials, except for the inserted gap. The frequency of the background used to carry the gap was systematically varied throughout the remainder of the session. The background noise consisted of bandpass-filtered noise bands at four frequencies (10, 12, 16, and 24 kHz) and also BBN presented with and without gaps. Stimuli were presented in ascending order from 10 kHz to BBN for each test. The offset of the gap began 50 ms before the onset of the startle stimulus (50 ms duration, 1 ms rise/fall time). In conditions where PPI is measured, prepulse stimuli match the backgrounds used in Gap detection in both spectral and intensity dimensions. PPI testing is essentially the inverse of Gap detection testing and serves as a valuable control for hearing loss and temporal deficits that might explain any Gap detection deficits. For example, rather than presenting a 60 dB, 10 kHz signal in the background continuously and embedding a 50-ms Gap before startling the animal, during PPI testing the background is quiet and a 50-ms duration, 60 dB, 10 kHz prepulse signal is presented before startling the animal.

Inclusion Phase of GPIAS Groups

Baseline auditory brainstem responses (ABR) were measured from both ears (eight animals by group) using tone burst presented at 2, 4, 8, 16, and 25 kHz; tone bursts intensity varied from 90 to 0 dB SPL at each frequency. The measures were performed under isoflurane anesthesia. One animal with a threshold above 40 dB SPL in one ear at 16 kHz was considered to have preexisting hearing loss and was eliminated from the study. The animals selected were randomly distributed to the six groups (three treatments groups each with an experimental and control group).

GPIAS testing ($n = 6$ animals per group) was carried at four frequencies, 10, 12, 16, and 24 kHz and BBN. Background sounds in the startle chamber consist of a 60-dB peak SPL level broadband noise (BBN). Animals were acclimated to the apparatus for 2 consecutive days the week before the first GAP baseline measure was collected and baseline GPIAS measures were obtained. During baseline GPIAS tests, the amplitude of the startle reflex on trials containing the gap insertion (Gap) was expected to be lower than those without the gap (No Gap). Animals with puta-tive tinnitus with features similar to the background sound fail to detect the silent gap, presumably due to their tinnitus.

Therefore, acoustic startle amplitudes on gap trials were presumably similar to those without the gap (i.e., little or no inhibition of the startle reflex). For each ani-mal, the calculation was realized from the median value (median ± median absolute deviation) of the 12 trials (of fewer because of aberrant trials) for Gap detection test (No Gap, Gap). In addition, for each rat, the percentage of Gap inhibition (($1 - $ Gap)/ No Gap) × 100 was calculated and the obtained values were average for each group (mean ± SEM). As the startle occurred at 250 ms, only pairs of trials of No PPI/PPI − No Gap/Gap exhibiting a startle response within the time interval from 265 to 320 ms (i.e., 15–70 ms after startle stimulus) were included.

MEMRI Analysis Methods

The MEMRI analysis is based on three different processing methods of the same set of MR images and regions of interest (ROIs), namely signal-to-noise ratio (SNR), signal intensity ratio (SIR), and the ($\Delta R2/R2$) methods. The SNR is defined as the mean MRI signal intensity in a given region divided by the image noise intensity [43, 44] (Fig. 8).

Fig. 8 $T1$-weighted MR coronal brain slice acquired with 9.4 T scanner (400 MHz). Arrows point to examined regions of interest (ROIs): Image noise, superficial temporal muscle (STM), inferior colliculus (auditory structure), and cerebral cortex (nonauditory structure)

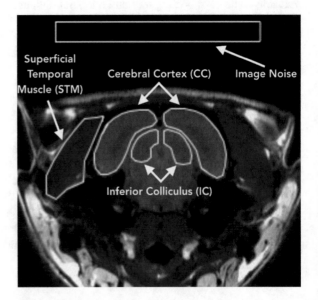

The SIR is defined as the mean MRI signal intensity in a given region of the brain divided by the mean MRI signal intensity in the superficial temporal muscle [7, 35, 48]. Finally, the $\Delta R2/R2$ method is based on the ratio enhancement of the relaxation rate $R2$ between two regions of interest: the cerebral cortex (visual cortex, entorhinal cortex, subiculum, and cingulum areas) and in the auditory structures (i.e., the IC). Figure 9 presents the procedures used for acquiring and processing the MR images and analyzing the results.

One axial brain slice that included the IC is selected. From a homebuilt program using a Matlab interface, the ROIs that included: (1) the superficial temporal muscle (STM), (2) image noise intensity, (3) IC and (4) cerebral cortex (CC) are segmented as illustrated in Fig. 8. The ROI segmentations of the IC and CC were carried out on both left and right sides of the brain with administrations of $MnCl_2$ by the intraperitoneal and transtympanic ipsilateral side routes.

$R2$ relaxation rate maps can be calculated voxel by voxel in each ROI according to the following expression:

$$R2 = \frac{1}{TE} \times \log \frac{echo1}{echo2}$$

as presented using a red, green, blue (RGB) color mapping in Fig. 10a, b.

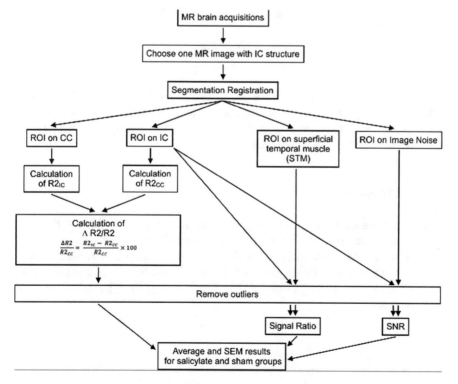

Fig. 9 Schematic representation of MR images processing protocol for IC, CC, image noise, and STM ROIs extraction. Three different analysis methods are displayed: signal-to-noise ratio (SNR), signal intensity ratio (SIR), and ($\Delta R2/R2$) method

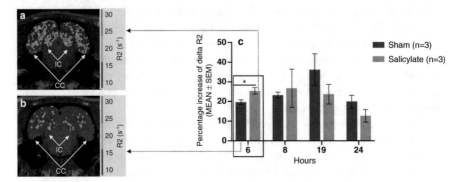

Fig. 10 (a) $R2$ mapping (RGB colormap) superimposed over $T1$-weighted MR images on IC and CC of both sides of image. MR images were acquired 6 h after intraperitoneal Mn^{2+} injection for salicylate induced tinnitus rat. (**b**) MR image 6 h after administration of IP administration of $MnCl_2$ for sham rat without treatment. (**c**) Measurements $\Delta R2/R2$ in IC at 6, 8, 19, and 24 h after $MnCl_2$ administration. $*p < 0.05$

To rule out interindividual $MnCl_2$ diffusion problems especially by the TT route [44], the percentage increase of $\Delta R2/R2$ was calculated from the $R2$ relaxation rate on the IC and CC according to the following equation:

$$\frac{\Delta R2}{R2_{CC}} = \frac{R2_{IC} - R2_{CC}}{R2_{CC}} \times 100$$

Figure 11 presents an example of the distribution of $\Delta R2/R2$ values for both groups (sham and salicylate) showing the typical and atypical $\Delta R2/R2$ values to remove outliers that are far from the main part of the distribution. Finally, the SNR and SIR analyses were performed on the selected animals.

Statistical Analysis

Descriptive statistics by groups are provided as the mean ± SEM (arithmetic mean ± standard error of the mean) for SNR quantification, SIR, and percentage increase using the $\Delta R2$ methods. Statistical significance was determined using Student's t-test using Prism software with the following significance value: $*p < 0.05$, $**p < 0.01$, and $***p < 0.001$; n represents the number of independent experiments.

Fig. 11 Example showing the distribution of $\Delta R2/R2$ values for salicylate and sham groups. Data too far from all data in the group (close data) were removed

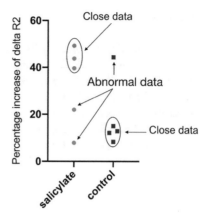

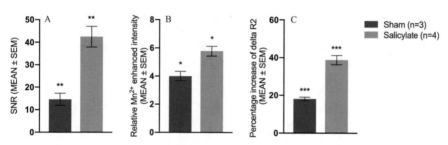

Fig. 12 Comparison between salicylate-treated group and sham group with SNR analysis method (**a**), SIR method (**b**), and $\Delta R2/R2$ method (**c**). $*p < 0.005$, $**p < 0.01$, $***p < 0.001$

3.3 Results

Series A

For the series A experiments, rats received 20 μL of $MnCl_2$ by the TT route 24 h before the MRI session. The SIT group received sodium salicylate (300 mg/kg, i.p.) once a day from D1 to D4 and the sham group was injected with a similar volume of saline solution. The ipsilateral IC showed an uptake of manganese after salicylate exposure when the contrast agent was injected 24 h prior to imaging and when the last dose of sodium salicylate was injected 18 h before MEMRI. As shown in Fig. 12, a significant increase of SIR, SNR, and $\Delta R2/R2$ was observed in the SIT group. These results confirmed an increase of neural activity in the IC under salicylate treatment using a drug dose known to consistently induce tinnitus [7, 43, 48, 51]. A lower increase was observed when the MR images were analyzed with the SIR method compared to $\Delta R2/R2$ and SNR methods. In addition, SNR results showed the largest mean difference between both groups (sham and salicylate), but we observed less variability of statistical significance with the $\Delta R2/R2$ analysis method.

Series B

All animals injected with $MnCl_2$ by intraperitoneal administration at 6, 8, 19, or 24 h prior to imaging did not exhibit significant differences in MEMRI results with the SNR and SIR analysis methods ($p > 0.05$). With the $\Delta R2/R2$ analysis method, the group that received an injection at 6 h before the MEMRI showed a significant difference between the salicylate treated group and the sham group ($p = 0.047$) (Fig. 10c). The $\Delta R2/R2$ mappings in Fig. 10a, b revealed higher $\Delta R2/R2$ values on ROIs for the IC and CC for the salicylate-induced tinnitus group compared to the sham group.

Series C

To validate MEMRI analysis methods and confirm the presence of tinnitus, we added the GPIAS behavior data for the same animals. In Fig. 13, GPIAS was applied at baseline and 2 h after the first salicylate administration. The rats, which fulfilled the inclusion criteria at baseline measures (with ABR $\leq$40 dB at 16 kHz in both

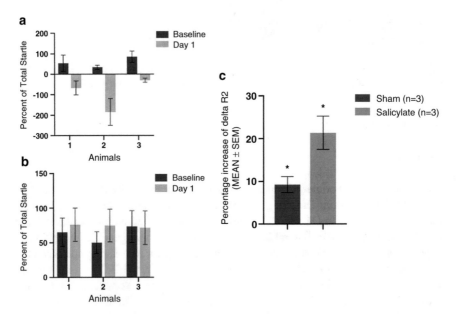

Fig. 13 Relationship between GAP data and Delta $R2$ method. (**a**) GAP detection before and after sodium salicylate administration (300 mg/kg) for three salicylate-treated rats. After collecting the GAP data, the sham and salicylate-treated rats were assessed with MEMRI. (**b**) GAP detection during baseline and day 1 for three sham rats. (**c**) Percentage increase of delta $R2$ between sham group and salicylate-induced tinnitus group

ears, with a % of PPI inhibition at BBN ≥50% and with a % of Gap inhibition at BBN ≥20%), were selected and distributed to the two groups. Exclusions of animals from the analysis were performed based on three predefined criteria:

- The time of response to the pulse: from 265 to 320 ms
- The percentage of Gap inhibition <20% at baseline for all groups
- The percentage of PPI inhibition (previously defined) <50% at baseline for all groups

As a reminder, for all animals at baseline, the startle reflex with gap insertion (Gap) was expected to be significantly lower than without the gap (No Gap) because the gap was assumed to be perceptible in the absence of tinnitus. In contrast, the animals with putative tinnitus that presumably have tinnitus should have startle reflexes similar to the No Gap condition because the tinnitus fills in the silent gap. For the sham group, the nonresponsive animals at baseline (who have weak Gap responses) were excluded for the concerned frequency. At baseline, all groups presented a similar % of Gap inhibition. The Sham group had a constant % of Gap inhibition at D1. As expected, a decrease of the % of Gap inhibition was observed in the Salicylate group compared to the Sham group. After analysis of GPIAS results, six rats were selected based on the criteria described above to perform MEMRI on three animals per group: three salicylate-treated rats (Fig. 13a) and three sham rats (Fig. 13b).

At day 2, all animals were injected with $MnCl_2$ (i.p.) 6 h before and then with either salicylate (experimental group) or vehicle (control group) 4 h before MEMRI measures. No significant differences in manganese uptakes were found with SNR and SIR results with our experimental protocol. In accordance with GPIAS results, $\Delta R2/R2$ showed a significant difference between groups confirming the presence of tinnitus in the salicylate-treated animals in Fig. 13c.

3.4 Discussion

In the literature, an enhancement of the ^1H NMR signal in the presence of Mn^{2+} in the auditory pathway associated with noise or salicylate-induced tinnitus was observed [53]. To our knowledge, we report for the first time the use of three distinct MEMRI analysis methods to identify significant changes in Mn^{2+} on the same sham and salicylate-treated rats. If SNR and SIR analysis methods are well-established methods [7, 35, 43, 44, 54], the $\Delta R2/R2$ method is an innovative approach based on the NMR relaxation rate $R2$. In our case, the three MEMRI analysis methods were compared on the same tinnitus animal model (SIT model) in order to determine the optimal processing protocol to apply.

Comparison Between Three MEMRI Analysis Methods

To achieve our analysis data on the different experimental protocols (Figs. 10 and 12) in the first step, the $\Delta R2/R2$ method was applied enabling to determine a selection of animals after removing outliers $\Delta R2/R2$ values (example in Fig. 11). All rats with atypical $\Delta R2/R2$ values in IC were excluded from the animal cohort. Thus, SNR and SIR were then carried out. Figure 12 presents data of MEMRI analysis methods from MRI acquisitions of the rat brain 24 h after $MnCl_2$ TT administration on a SIT model. Significant increases of Mn^{2+} signal intensities were observed in each analysis method. However, we observed lower variability and larger statistically significant values with the $\Delta R2/R2$ analysis method. This new method is based on an NMR relaxation rate $(R2)$ and is proportional to the $MnCl_2$ concentration as shown in the following equation:

$$Ri = Ri_0 + r_i \left(\left[MnCl_2 \right]_{\text{diffusion}} + \left[MnCl_2 \right]_{\text{tinnitus}} \right); \quad i = 1, 2$$

From the work of Caravan and colleagues [54], we estimated the relaxivity $(r_1 = 5.14 \text{ mM}^{-1}; r_2 = 117 \text{ mM}^{-1})$ and found agreement with their data. The $\Delta R2/R2$ method relies on the NMR transversal relaxation rate $(R2)$ changes, while SNR and SIR are based on longitudinal relaxation rate $R1$ contrasts from the signal intensities in IC and cerebral cortex (CC). In our study, we observed that the amount of Mn^{2+} uptake into the brain can vary between rats, particularly during TT administration making the SNR and SIR analysis difficult to perform. This interindividual diffusion of $MnCl_2$ is corrected in the $\Delta R2/R2$ approach between the IC and CC. In series A, for the rest of the statistical analysis, three salicylate-treated rats and two sham rats were excluded because of atypical $\Delta R2/R2$ values. It has been shown in the literature that strong enhancement of signal intensity after $MnCl_2$ administration was observed in some nonauditory areas like the hippocampus [8, 35, 37]. Preliminary data demonstrated that for two SIT excluded rats, the mean relaxation rate $(R2)$ in hippocampus was clearly similar to mean $R2$ values in the hippocampus of rats, which received no $MnCl_2$ or salicylate. Based on these results, it appears that there is considerable variability in $MnCl_2$ administration and uptake. For the last SIT rat excluded, it may be that the rat did not react to the sodium salicylate drug.

Comparing all analysis techniques, the $\Delta R2/R2$ method shows less variable results in the SIT and sham groups and can provide robust results compared to the SNR and SIR methods. This $\Delta R2/R2$ method is based on a semiautomatic technique decreasing human bias and reducing potential human errors. In addition, $R2$ is proportional to the $MnCl_2$ concentration, so the $\Delta R2/R2$ method could in principle quantify the presence of $MnCl_2$ and identifying hyperactivity due to tinnitus with a salicylate-induced tinnitus model.

Transtympanic (TT) vs. Intraperitoneal (IP) Mn²⁺ Administration

In the SIT model, TT and IP administration of $MnCl_2$ revealed significant differences between the salicylate-treated and sham groups. Unfortunately, the local $MnCl_2$ administration has some negative consequences in terms of animal wellness with a large variability of $MnCl_2$ diffusion into the cochlea. As previously discussed this could result from position of the needle during TT injections [44]. Afterward, we tested IP administration at several injection times in order to limit side effects. From the MEMRI study and post processing analysis, we obtained significant results only from the $\Delta R2/R2$ method at 6-h posttreatment (Fig. 10c). It has to be noted that no animal was excluded using $\Delta R2/R2$ method. Conversely, for TT administration, significant results were obtained with the different analysis methods. Moreover, the SIT results with the percentage increase of $\Delta R2/R2$ were higher with the TT administration than IP administration (Figs. 10 and 12c; $p < 0.001$ vs. $p < 0.05$). IP administration provides a lower and bilaterally homogeneous concentration of Mn^{2+} compared to TT administration.

GPAIS vs. MEMRI

We compared our GPIAS with MEMRI results obtained by IP administration. Despite a less significant difference with $\Delta R2/R2$ analysis method obtained with $MnCl_2$ administration (IP) at 6 h before MEMRI measures, our GPIAS results indicative of tinnitus were associated with an increase in $\Delta R2/R2$ in the IC suggesting that this MEMRI methods may be adequate to detect a tinnitus-like neural signature with a salicylate-induced tinnitus model (Fig. 13).

4 Conclusion

In summary, compared to the SNR and SIR analytical methods, the delta $\Delta R2/R2$ method shows less variability and higher level of statistical significance when assessing Mn^{2+} in the IC when comparing salicylate-treated rats with sham. The method is based on a semiautomatic segmentation of the brain, which reduces experimental bias and increases consistency thereby eliminating operator-dependent issues. The $R2$ relaxation rate is proportional to $MnCl_2$ concentration, which makes it sensitive and quantitative to the accumulation of Mn^{2+}. Therefore, this new innovative method of quantification is potentially more robust and reproducible compared to other MEMRI analysis methods. If one assumes that Mn^{2+} uptake in the IC is a reliable metric of tinnitus with all forms of this disorder, then the $\Delta R2/R2$ method could conceivably be used to evaluate the effectiveness of some medicines to reduce tinnitus.

References

1. Cunningham LL, Tucci DL (2017) Hearing loss in adults. N Engl J Med 377:2465–2473
2. Wilson BS, Tucci DL, Merson MH, O'Donoghue GM (2017) Global hearing health care: new findings and perspectives. Lancet 390:2503–2515
3. Sultana H, Mumtaz N, Dawood T (2018) Type and degree of hearing loss in patients with tinnitus. International Journal of Rehabilitation Sciences (IJRS) 7(01):24–27
4. Basile CE, Fournier P, Hutchins S, Hebert S (2013) Psychoacoustic assessment to improve tinnitus diagnosis. PLoS One 8:e82995
5. Stolzberg D, Salvi RJ, Allman BL (2012) Salicylate toxicity model of tinnitus. Front Syst Neurosci 6:28
6. Brozoski TJ, Ciobanu L, Bauer CA (2007) Central neural activity in rats with tinnitus evaluated with manganese-enhanced magnetic resonance imaging (MEMRI). Hear Res 228:168–179
7. Holt AG, Bissig D, Mirza N, Rajah G, Berkowitz B (2010) Evidence of key tinnitus-related brain regions documented by a unique combination of manganese-enhanced MRI and acoustic startle reflex testing. PLoS One 5:e14260
8. Yu X, Wadghiri YZ, Sanes DH, Turnbull DH (2005) In vivo auditory brain mapping in mice with Mn-enhanced MRI. Nat Neurosci 8:961–968
9. Chen YC, Li X, Liu L, Wang J, Lu CQ, Yang M, Jiao Y, Zang FC, Radziwon K, Chen GD et al (2015) Tinnitus and hyperacusis involve hyperactivity and enhanced connectivity in auditory-limbic-arousal-cerebellar network. elife 4:e06576
10. Harding GW, Bohne BA (2009) Relation of focal hair-cell lesions to noise-exposure parameters from a 4- or a 0.5-kHz octave band of noise. Hear Res 254:54–63
11. Harding GW, Bohne BA, Ahmad M (2002) DPOAE level shifts and ABR threshold shifts compared to detailed analysis of histopathological damage from noise. Hear Res 174:158–171
12. Gauvin DV, Yoder J, Koch A, Zimmermann ZJ, Tapp RL (2017) Down for the count: the critical endpoint in ototoxicity remains the cytocochleogram. J Pharmacol Toxicol Methods 88:123–129
13. Viberg A, Canlon B (2004) The guide to plotting a cochleogram. Hear Res 197:1–10
14. Neal C, Kennon-McGill S, Freemyer A, Shum A, Staecker H, Durham D (2015) Hair cell counts in a rat model of sound damage: effects of tissue preparation & identification of regions of hair cell loss. Hear Res 328:120–132
15. Bohne BA (1972) Location of small cochlear lesions by phase contrast microscopy prior to thin sectioning. Laryngoscope 82:1–16
16. Hardie NA, MacDonald G, Rubel EW (2004) A new method for imaging and 3D reconstruction of mammalian cochlea by fluorescent confocal microscopy. Brain Res 1000:200–210
17. Bohne BA, Harding GW (2011) Microscopic anatomy of the mouse inner ear, laboratory manual, 3rd edition. Washington University Press, pp 1–2, St. Louis, MO, USA
18. Ehret G, Frankenreiter M (1977) Quantitative analysis of cochlear structures in the house mouse in relation to mechanisms of acoustical information processing. J Comp Physiol 122:65–85
19. Keithley EM, Feldman ML (1982) Hair cell counts in an age-graded series of rat cochleas. Hear Res 8:249–262
20. Spoendlin H, Brun JP (1974) The block-surface technique for evaluation of cochlear pathology. Arch Otorhinolaryngol 208:137–145
21. Akil O, Seal RP, Burke K, Wang C, Alemi A, During M, Edwards RH, Lustig LR (2012) Restoration of hearing in the VGLUT3 knockout mouse using virally mediated gene therapy. Neuron 75:283–293
22. Bohne BA, Kimlinger M, Harding GW (2017) Time course of organ of Corti degeneration after noise exposure. Hear Res 344:158–169
23. Chen GD, Decker B, Krishnan Muthaiah VP, Sheppard A, Salvi R (2014) Prolonged noise exposure-induced auditory threshold shifts in rats. Hear Res 317(1–8):1

24. Takada Y, Takada T, Lee MY, Swiderski DL, Kabara LL, Dolan DF, Raphael Y (2015) Ototoxicity-induced loss of hearing and inner hair cells is attenuated by HSP70 gene transfer. Mol Ther Methods Clin Dev 2:15019
25. Akil O, Lustig LR (2013) Mouse cochlear whole mount immunofluorescence. Bio Protoc 3:e332
26. Kaiming H, Xiangyu Z, Shaoqing R, Jian S (2015). Deep Residual Learning for Image Recognition. 2016 IEEE Conference on Computer Vision and Pattern Recognition (CVPR): 770–778
27. Sinno P, Qiang, Y (2010). A Survey on Transfer Learning. Knowledge and Data Engineering, IEEE Transactions on. 22. 1345 - 1359. doi: 10.1109/TKDE.2009.191
28. Monro HRS (1951) A stochastic approximation method. Ann Math Stat 22:400–407
29. Ding D, Jiang H, Chen GD, Longo-Guess C, Muthaiah VP, Tian C, Sheppard A, Salvi R, Johnson KR (2016) N-acetyl-cysteine prevents age-related hearing loss and the progressive loss of inner hair cells in gamma-glutamyl transferase 1 deficient mice. Aging (Albany NY) 8:730–750
30. Sanz L, Murillo-Cuesta S, Cobo P, Cediel-Algovia R, Contreras J, Rivera T, Varela-Nieto I, Avendano C (2015) Swept-sine noise-induced damage as a hearing loss model for preclinical assays. Front Aging Neurosci 7:7
31. Lockwood AH, Salvi RJ, Burkard RF (2002) Tinnitus. N Engl J Med 347:904–910
32. Jastreboff PJ, Brennan JF, Coleman JK, Sasaki CT (1988) Phantom auditory sensation in rats: an animal model for tinnitus. Behav Neurosci 102:811–822
33. Boyen K, Baskent D, van Dijk P (2015) The gap detection test: can it be used to diagnose tinnitus? Ear Hear 36:e138–e145
34. Turner JG, Brozoski TJ, Bauer CA, Parrish JL, Myers K, Hughes LF, Caspary DM (2006) Gap detection deficits in rats with tinnitus: a potential novel screening tool. Behav Neurosci 120:188–195
35. Muca A, Standafer E, Apawu AK, Ahmad F, Ghoddoussi F, Hali M, Warila J, Berkowitz BA, Holt AG (2018) Tinnitus and temporary hearing loss result in differential noise-induced spatial reorganization of brain activity. Brain Struct Funct 223:2343–2360
36. Galazyuk A, Hebert S (2015) Gap-prepulse inhibition of the acoustic startle reflex (GPIAS) for tinnitus assessment: current status and future directions. Front Neurol 6:88
37. Lee JW, Park JA, Lee JJ, Bae SJ, Lee SH, Jung JC, Kim MN, Lee J, Woo S, Chang Y (2007) Manganese-enhanced auditory tract-tracing MRI with cochlear injection. Magn Reson Imaging 25:652–656
38. Pautler RG (2004) In vivo, trans-synaptic tract-tracing utilizing manganese-enhanced magnetic resonance imaging (MEMRI). NMR Biomed 17:595–601
39. Cheung SW, Nagarajan SS, Bedenbaugh PH, Schreiner CE, Wang X, Wong A (2001) Auditory cortical neuron response differences under isoflurane versus pentobarbital anesthesia. Hear Res 156:115–127
40. Malheiros JM, Paiva FF, Longo BM, Hamani C, Covolan L (2015) Manganese-enhanced MRI: biological applications in neuroscience. Front Neurol 6:161
41. Schumacher JW, Schneider DM, Woolley SM (2011) Anesthetic state modulates excitability but not spectral tuning or neural discrimination in single auditory midbrain neurons. J Neurophysiol 106:500–514
42. Ter-Mikaelian M, Sanes DH, Semple MN (2007) Transformation of temporal properties between auditory midbrain and cortex in the awake Mongolian gerbil. J Neurosci 27:6091–6102
43. Jin SU, Lee JJ, Hong KS, Han M, Park JW, Lee HJ, Lee S, Lee KY, Shin KM, Cho JH et al (2013) Intratympanic manganese administration revealed sound intensity and frequency dependent functional activity in rat auditory pathway. Magn Reson Imaging 31:1143–1149
44. Lee HJ, Yoo SJ, Lee S, Song HJ, Huh MI, Jin SU, Lee KY, Lee J, Cho JH, Chang Y (2012) Functional activity mapping of rat auditory pathway after intratympanic manganese administration. Neuroimage 60:1046–1054

45. Paul AK, Lobarinas E, Simmons R, Wack D, Luisi JC, Spernyak J, Mazurchuk R, Abdel-Nabi H, Salvi R (2009) Metabolic imaging of rat brain during pharmacologically-induced tinnitus. Neuroimage 44:312–318
46. Rancz EA, Moya J, Drawitsch F, Brichta AM, Canals S, Margrie TW (2015) Widespread vestibular activation of the rodent cortex. J Neurosci 35:5926–5934
47. Silva AC, Bock NA (2008) Manganese-enhanced MRI: an exceptional tool in translational neuroimaging. Schizophr Bull 34:595–604
48. Groschel M, Gotze R, Muller S, Ernst A, Basta D (2016) Central nervous activity upon systemic salicylate application in animals with kanamycin-induced hearing loss—a manganese-enhanced MRI (MEMRI) study. PLoS One 11:e0153386
49. Jung DJ, Han M, Jin SU, Lee SH, Park I, Cho HJ, Kwon TJ, Lee HJ, Cho JH, Lee KY et al (2014) Functional mapping of the auditory tract in rodent tinnitus model using manganese-enhanced magnetic resonance imaging. Neuroimage 100:642–649
50. Liu Y, Li X, Ma C, Liu J, Lu H (2005) Salicylate blocks L-type calcium channels in rat inferior colliculus neurons. Hear Res 205:271–276
51. Hu SS, Mei L, Chen JY, Huang ZW, Wu H (2014) Expression of immediate-early genes in the inferior colliculus and auditory cortex in salicylate-induced tinnitus in rat. Eur J Histochem 58:2294
52. Spivak M, Weston J, Bottou L, Kall L, Noble WS (2009) Improvements to the percolator algorithm for peptide identification from shotgun proteomics data sets. J Proteome Res 8:3737–3745
53. Bauer CA, Brozoski TJ, Rojas R, Boley J, Wyder M (1999) Behavioral model of chronic tinnitus in rats. Otolaryngol Head Neck Surg 121:457–462
54. Caravan P, Farrar CT, Frullano L, Uppal R (2009) Influence of molecular parameters and increasing magnetic field strength on relaxivity of gadolinium- and manganese-based T1 contrast agents. Contrast Media Mol Imaging 4:89–100

Printed in the United States
by Baker & Taylor Publisher Services